MEDICINE

IN THE NEWS

SCIENCE NEWS FLASH

The Environment inthe News
Genetics in the News
Medicine in the News
Water in the News

MEDICINE

IN THE NEWS

BRIAN R. SHMAEFSKY, PH.D.

YAEL CALHOUN
Consulting Editor

ON THE COVER: A surgeon operates on a patient's heart using the da Vinci surgical robot.

Medicine in the News

Chelsea House
An imprint of Infobase Publishing
132 West 31st Street
New York NY 10001

Library of Congress Cataloging-in-Publication Data
Shmaefsky, Brian.
Medicine in the news / Brian R. Shmaefsky.
p. cm. — (Science news flash)
Includes bibliographical references and index.
ISBN-13: 978-0-7910-9256-9 (hardcover)
ISBN-10: 0-7910-9256-9 (hardcover)
1. Genetic engineering. 2. Medical technology. 3. Medical genetics.
4. Biotechnology. I. Title. II. Series.
QH442.S52 2007
610—dc22 2007000183

Chelsea House books are available at special discounts when purchased in bulk quantities for businesses, associations, institutions, or sales promotions. Please call our special sales department in New York at (212) 967-8800 or (800) 322-8755.

You can find Chelsea House on the world wide web at http://www.chelseahouse.com

Series design by Annie O'Donnell
Cover design by Ben Peterson

Printed in the United States of America

Bang EJB 10 9 8 7 6 5 4 3 2 1

This book is printed on acid-free paper.

Contents

I. Stem Cells and Cloning 1

What's in the News? 2

Introduction 3

What Are Stem Cells? 6

Current and Future Uses of Stem Cells 11

What Is Cloning? 17

Current and Future Uses of Cloning 21

References 24

Further Reading 25

II. Gene Therapies and Tissue Engineering 27

What's in the News? 28

Introduction 30

What Is Gene Therapy? 31

Current and Future Uses of Gene Therapy 35

What Is Tissue Engineering? 39

Current and Future Uses of Tissue Engineering 42

References 51

Further Reading 52

III. Bionanotechnology and Robotics 55

What's in the News? 56

Introduction 58

What Is Bionanotechnology? 60
Current and Future Uses of Bionanotechnology 63
What Is Robotics? 67
Current and Future Uses of Robotics 70
References 77
Further Reading 78

IV. Pharmacogenetics and Precision Vaccines 79
What's in the News? 80
Introduction 81
What Is Pharmacogenetics? 83
Current and Future Uses of Pharmacogenetics 85
What Are Precision Vaccines? 88
Current and Future Uses of Precision Vaccines 91
References 94
Further Reading 96

V. Medical Instrumentation 97
What's in the News? 98
Introduction 101
What Are Medical Instruments? 103
Future Types of Medical Instruments 110
References 116
Further Reading 117

Glossary 118
Index 133

Stem Cells and Cloning

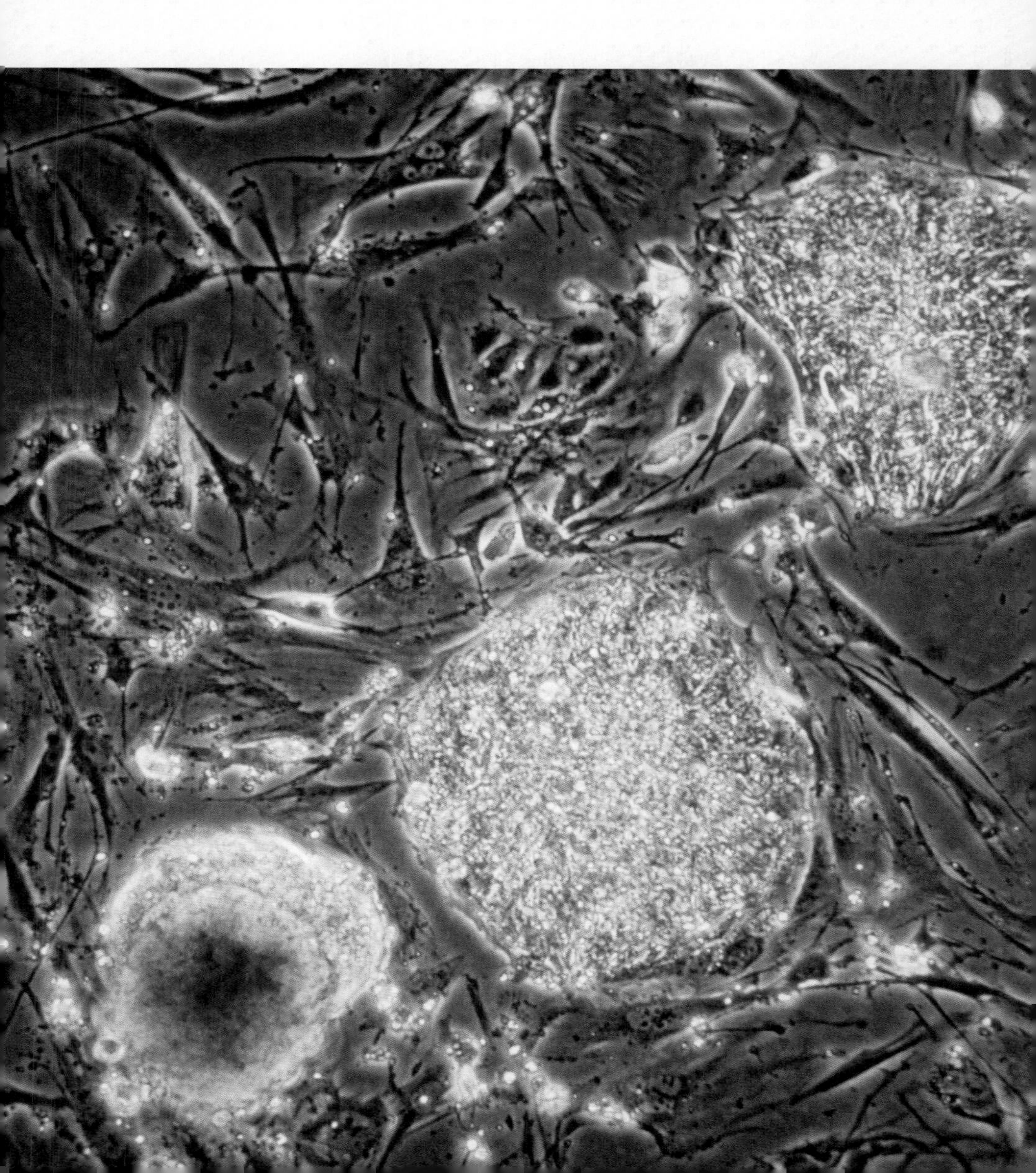

WHAT'S IN THE NEWS?

Spinal cord damage has been the bane of the medical community. According to the National Spinal Cord Injury Statistical Center, each year in the United States, 11,000 new cases of spinal cord damage occur. Actor Christopher Reeve brought spinal cord injury public attention when he was paralyzed as a result of a horse riding accident in 1995. Even with modern surgical practices, little can be done to repair an injured spinal cord. Scientists have been unable to repair the many thousands of nerve pathways that direct information back and forth between the brain and the body. Over time, the condition of an injured spinal cord deteriorates as **nerve cells** next to the injured nerves die off progressively, leaving behind further loss of feeling or movement.

The frustrations of using surgery for correcting spinal cord damage led to a new way of viewing the treatment of nervous system disorders. In 1995, a team of researchers working on a cooperative project between the University of Miami School of Medicine and the Centre Hospitalier Universitaire Vaudois, in Lausanne, Switzerland, tried a novel way of encouraging nervous system repair.[1] They used special chemicals called **growth factors** to stimulate healing of the damaged spinal cords in rats. The chemicals they used were brain-derived neurotrophic factor (BDNF) and neurotrophin-3 (NT-3). These two growth factors promote action of cells involved in the healing and development of nerves. The research resulted in some nerve **regeneration**. However, the repair was not extensive enough to ensure complete healing of major spinal injuries.

Research carried out in 2006 at Washington University School of Medicine in St. Louis, Missouri, restored leg movement in injured rats by using stem cells to replace damaged body parts. The researchers transplanted mouse embyro stem cells into an injured area of a rat's spine nine days after the rat received damage to the spine.[2] As in the 1995 experiment, growth factors were used to make sure that the stem cells grew into nerve cells and other spinal cord cells. The rats regained much of the use of their legs within two weeks after the treatment. This study was made possible by the many stem cell investigations carried on throughout the world since the late 1990s. Stem cell research is continuously in the news as a potential treatment for a variety of human ailments.

INTRODUCTION

The terms **clone** and **stem cell** are very likely to be encountered in almost every current news report about advances in disease therapy. A clone describes an identical copy of a cell or an organism. Stem cells are unspecialized cells that can develop into different types of cells. Cloned cells that were modified to produce a particular drug have been used to produce a wide variety of medicines. Almost all modern vaccines are being produced this way. Stem cell technology involves the use of cloned cells or body structures. They are useful in medicine in a variety of ways ranging from the treatment of **genetic disorders** to the regeneration of damaged **organs**. Innovations using cloned cells and stem cells have many future benefits that may eliminate much of the suffering from disease and injury. However, there are potential health risks with using the technology as well as serious ethical issues associated with stem cell procurement.

Cell Basics

Cells are the most basic unit of living structure for most organisms (Figure 1.1). Scientists divide cells into three major parts: **cell membrane**, **cytoplasm**, and **genome**. The cell membrane is a lipid and protein covering that encloses the cell allowing for the exchange of materials into and out of the cell. The cytoplasm makes up the structures within the cell membrane. In simple cells such as those found in **prokaryotes**, the cytoplasm is a fluid that carries out most cell functions. Prokaryotes are single-celled organisms, such as bacteria, that lack a compartmentalized

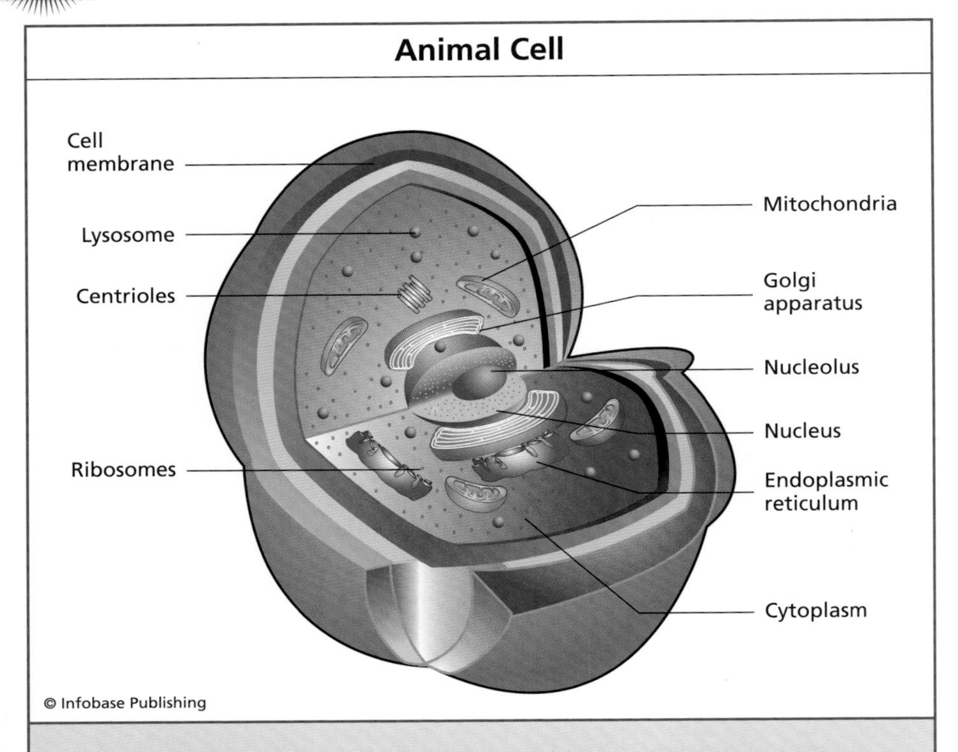

Figure 1.1 The cell is the basic unit of life. Structures called organelles help cells carry out their living functions.

cytoplasm. Complex organisms such as **eukaryotes** have a cytoplasm containing **organelles**. Organelles are small compartments that carry specific features of a cell's function. Animals and plants have eukaryotic cells.

Eukaryotes contain two major categories of organelles. One group is called the **endomembrane system** organelles and the other is called the **endosymbiont** organelles. There are five common types of organelles in the endomembrane system: the **nuclear membrane**, **endoplasmic reticulum**, **Golgi apparatus**, **vesicle**, and cell membrane. These cell components transfer chemicals and cell parts to each other through a system of membrane tubes and through small sacks called **transport vesicles**. The nuclear membrane is responsible for passing information back and forth between the genome and the cytoplasm. Attached to the nuclear membrane is the endoplasmic reticulum that is responsible for the production of the fats and proteins needed for cell functions and structures. Ribosomes are structures found in the endoplasmic reticulum and are responsible for building proteins. The Golgi apparatus modifies, stores, and transports chemicals made in the endoplasmic reticulum. Sacks called transport vesicles move chemicals from the endoplasmic reticulum to the Golgi apparatus. A specialized vesicle called the **lysosome** contains chemicals capable of breaking down the cell components. It is involved in destroying the cell when the cell is injured or dies.

Endosymbiont organelles are prokaryotic organisms that live within the eukaryotic cell. The endosymbionts are passed on to offspring through the egg's cytoplasm. Endosymbiont organelles contain a genome that controls specific aspects of their life processes. Two major endosymbiont organelles are **mitochondria** and **chloroplasts**. Mitochondria carry out a bulk of a cell's **metabolism** (chemical processes). Chloroplasts are plant endosymbionts that perform a series of chemical reactions called **photosynthesis**. Photosynthesis uses sunlight to convert carbon dioxide and water into the compounds a plant needs for its functions and structures.

The term *genome* refers to the complete material passed down from one generation to the next. All cells use **deoxyribonucleic acid**, or **DNA**, as the heritable material. Eukaryotic cells have the genome located in a saclike structure called the **nucleus**. An organism's genome serves as a blueprint for the chemistry and organization of each cell. The genetic code is the basis of DNA information. Information provided by the genome is organized in information units called **genes**. A gene is commonly defined as a functional unit of heredity. The typical gene is a section of DNA located in a specific spot on the genome. A **chromosome** is made of **chromatin**, which is a threadlike collection of genes and other DNA. Many genes are arranged in clusters that program for specific characteristics of an organism.

WHAT ARE STEM CELLS?

Complex organisms such as animals and plants are made up of many types of cells — such as blood cells, nerve (neurons) cells, and bone cells. All cells in a body develop from the original undifferentiated cells that first grow when a sperm fertilizes an egg. Over time, cells become **differentiated**, or grow into bone, nerve, or blood cells, as they respond to various chemical factors in the body. Scientists do not fully understand all the mechanisms that cause cells to differentiate.

A stem cell is defined as a cell from which other types of cells of the body can develop. Scientists who studied the development of animals and plants in the late 1800s first discovered stem cells in **embryos**.[3] They recognized that certain cells were responsible for the production of specific types of tissues and body organs. The first stem cells were isolated in the early 1900s from the center of bones. These stem cells produced the different types of blood cells. In 1928, a German scientist named Hans Spemann discovered that a cell's nucleus controlled the division and differentiation of stem cells. By the 1950s, scientists recognized four major types of stem cells: **totipotential**, **pluripotential**, **multipotential**, and **unipotential**.

Types of Stem Cells

Totipotential stem cells are cells capable of producing any type of cell in an organism. They can also form a whole organism. A fertilized egg is totipotential. At one time it was believed that totipotential cells were present only in early stages of embryonic development in animals and humans. A study conducted in 2002 showed the presence of totipotential stem cells in adult humans. Dr. Catherine Verfaillie at the University of Minnesota discovered these cells in the **bone marrow** of adults. Bone marrow is responsible for the production of blood cells. Scientists at the **biotechnology** company MorphoGen Pharmaceuticals in San Diego found similar cells in mice. They discovered totipotential stems in adult mouse muscle and skin.

Pluripotential stem cells can develop into a wide array of cells that form major components of tissues in an organ. A cell called the **hematocytoblast** is a pluripotential stem cell that forms other stem cells. These stem cells in turn then form the various types of blood cells. Pluripotential cells are found in embryos and adults.

Multipotential cells are similar to pluripotential cells. They differ in that they only produce a limited variety of cell types found in a tissue (Figure 1.2). Many multipotential cells in humans are derived from a pluripotential cell. One type of multipotential cell in bone marrow is responsible for the formation of **red blood cells**. Red blood cells carry oxygen throughout the body.

Unipotential, or committed, stem cells are only able to produce one type of cell. They are very common in adults. These cells play an important role in growth and healing. Unipotential stem cells are often called tissue-specific cells because they form one particular type of tissue. One type of unipotential stem cell is responsible for producing the outer surface of the skin. The liver also has unipotential stem cells that normally sit inactive but have the potential to produce mature liver cells. Unipotential liver cells create new liver cells to replace those that have been damaged.

Control of Stem Cell Function

The differentiation of stem cells requires signals to switch genes on and off. These signals are called growth factors. There are a wide variety of growth factors ranging from chemicals produced by other cells to environmental features such as temperature, toxins, or the amount of oxygen in the fluids around a cell. The most common growth factors are categorized as **differentiation factors** and **mitogens**. Differentiation factors are chemicals that maintain the health of a cell and direct the differentiation of a stem cell. **Insulin-like Growth Factor-I** (IGF-I) is produced by a variety of cells. It assists other growth factors that direct the differentiation of blood cells from pluripotential and multipotential stem cells. **Epidermal growth factor** (EGF) is a chemical that binds the surface of unipotential skin cells. It stimulates the differentiation of skin stem cells into mature skin cells.

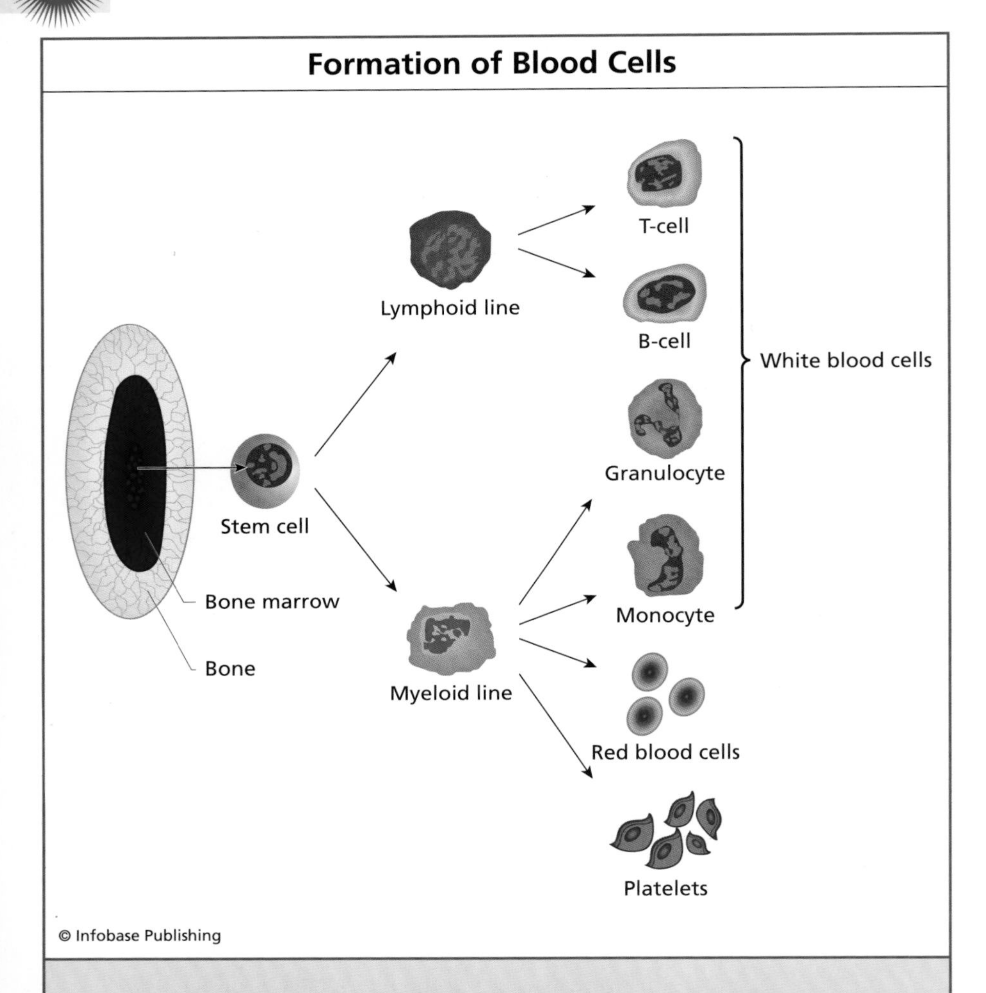

Figure 1.2 Stem cells are responsible for the production of other stem cells and for the formation of differentiated, mature cells. This chart illustrates the different types of blood cells that are generated by stem cells located in the bone marrow.

The body produces hundreds of differentiation factors that are released by various tissues and organs. Certain ones are produced only in an embryo. Many differentiation factors in adults are stimulated by specific signals produced during injury. **Estrogen** and **testosterone** are **hormones** that serve as differentiation factors during puberty. When acting as a differentiation factor, estrogen stimulates the development of secondary sexual characteristics in females. Testosterone initiates the formation of secondary sexual characteristics in males. Mitogens are not as diverse and numerous as differentiation factors. They primarily stimulate the replication of stem cells. A group of related mitogens called **endothelins** are responsible for skin cell replication.

Scientists are discovering overlapping roles between differentiation factors and mitogens. Many growth factors and mitogens take on interchangeable roles depending on the presence of other signals. A cell's contact with other cells is one important factor that modifies its response to growth factors and mitogens. For example, liver stem cells are more likely to form normal mature liver cells when in contact with adult liver cells. They are likely to produce abnormal cells if in contact with cells not associated with the liver. This feature is very important in determining the fate of cells in a developing embryo. It also helps organize the healing of damaged body organs in adults.

Sources of Stem Cells

Stem cells used in stem cell research and medical applications come from three major sources: embryos, **umbilical cords**, and adult tissues.[4] Embryonic stem cells are collected from early embryos that grew from artificially fertilized eggs (Figure 1.3). The embryos are cultured in a laboratory. Human embryonic stem cells are from discarded embryos produced in fertility clinics. The embryos are usually no more than five days old and will not survive further unless implanted in a woman. There are many ethical concerns about using human embryonic stem cells for research. Most of these concerns are focused on the fact that the harvesting of stem cells results in the destruction of the embryo. The American government and other countries have policies regulating and preventing the use of human embryonic stem cells for research and **therapeutic** applications. For example, current U.S. policy prohibits the use of federal funds for embryonic stem cell research.

Umbilical cord stem cells are collected from the narrow tube that connects the baby to the mother's body during pregnancy.

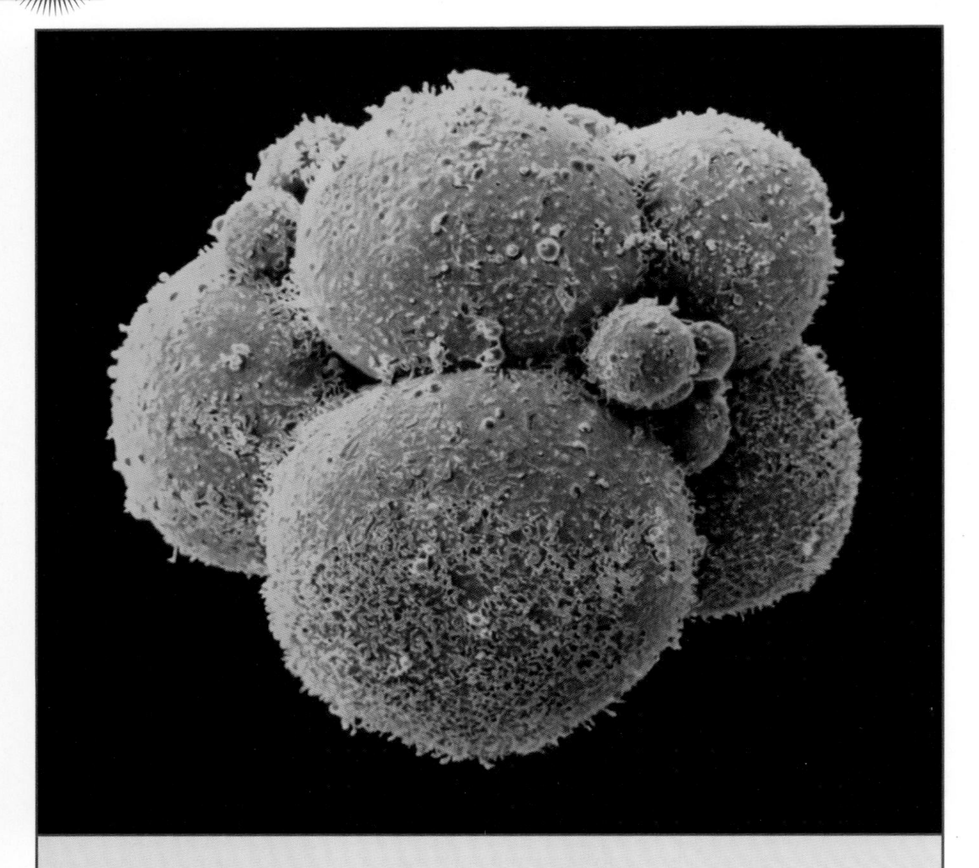

Figure 1.3 An eight-cell embryo, which contains embryonic stem cells, is shown above. Stem cells are believed to have the potential to treat many diseases, including Parkinson's disease, Alzheimer's disease, and diabetes.

The umbilical cord is rich in blood flow and is filled with a gel-like material that contains stem cells from the baby. Umbilical cords are normally discarded with the **placenta** after childbirth. The placenta is the structure that attaches the umbilical cord to the mother. Research studies carried out in 1988 showed that the umbilical cord was a valuable source of multipotential and pluripotential stem cells that develop into mature blood cells. Now it is known that other types of multipotential and pluripotential stem cells exist in the umbilical cord fluids. These cells can be differentiated into bone, liver, and nerve cells. There is strong evidence that totipotential stem cells from the infant and the mother are present in the umbilical cord.

Adult stem cells are usually obtained from body organs such as bone that regularly produce new cells. Certain organs such as skin that have a high capacity for regeneration are another potential source of stem cells. However, adult stem cells are usually unipotential and are very limited in their ability to differentiate into a variety of adult cell types. Multipotential and pluripotential adult stem cells are primarily found in bone marrow. Scientists have discovered two categories of adult stem cells in bone marrow: **hematopoietic stem cells** and **mesenchymal stem cells**. Hematopoietic stem cells are relatively simple to obtain from a person. They are removed using a sterile needle that siphons off some of the stem cells located in the hollow cavities within the bones of the arms, hips, and legs. These multipotential stem cells can be cultured **in vitro** to give rise to every blood cell type.

Mesenchymal stem cells are also collected from bone marrow. They are pluripotential cells capable of forming mature bone, cartilage, fat cells, ligament, muscle, tendon, and skin. It is believed that similar cells can be found in the muscles that move the skeleton.

CURRENT AND FUTURE USES OF STEM CELLS

The recent public aversion to stem cell research in the United States and other countries is a major concern for some researchers because of the many potential therapeutic uses of stem cells. Currently, embryonic stem cells are the most feasible stem cells for use in medicine. The need to obtain embryonic stem cells from human fetuses fuels much of the debate about stem cell research. Actors Christopher Reeve and Michael J. Fox both argued that one embryo could be used to help many lives affected by nervous system diseases and injuries.

Fetal cells are preferred over adult stem cells because adult cells have limited **plasticity**. The term *plasticity* is used to describe the ability of stem cells from an adult tissue to produce the differentiated cell types of another tissue. Scientists would prefer to work with stem cells that have high plasticity. Cells with high plasticity have a low degree of specialization and can be differentiated into any cell type. It is simpler to grow unspecialized cells in the laboratory because they have fewer specific needs for growth factors and nutrients. Plus, unspecialized stem cells can be induced to replicate new stem cells. Mature differentiated cells are difficult to keep alive in culture and usually do not produce new cells. It may take many years before researchers are able to induce the right amount of plasticity needed to make adult and umbilical stem cells as valuable as embryonic stem cells.

Stem Cell Lineages

Today, scientists have 60 existing human embryonic stem cell lineages available for research. These cells were produced in vitro with voluntary permission from donors.

Stem cell lineages are given names such as BF01, MI01, TE03, and WA04. The names are given by the National Institutes of Health and identify a particular lineage of cells developed by a research laboratory or pharmaceutical company. Letters represent the developer and numbers refer to the embryo. Laboratories throughout the world use these stem cells to carry out research for medical advances. Unfortunately, it is not possible to keep these 60 cell lineages healthy in culture indefinitely. New cells have to be collected from fresh embryos so that they are suitable for research and medical applications.[5]

Stem Cell Preparation

Stem cells are carefully tested and cultivated before being used for research or medical applications. The embryo donors and the embryo are tested to ensure the absence of genetic defects that may interfere with the use of the stem cells. Chemical testing is also done to evaluate the nature of the proteins on the surface of the stem cells. These proteins, called **antigens**, evaluate the ability of the cells to survive when placed in contact with other cells. Stem cells are ineffective if the proteins on the cells cause an **immune response** within the body of an individual receiving the cells. Upon passing the genetic and chemical testing, cells are removed from the embryo and placed in a sterile liquid **growth medium** for growth in a cell **incubator** (Figure 1.4). Growth medium is a complex mixture of chemicals that mimics the blood and the conditions found in the body.

Once in the growth medium, the stem cells are placed under a microscope so scientists can monitor growth and observe the DNA. Cultures with abnormal cells are discarded. Healthy cells are further tested in culture to see if they have any potential to form tumors or produce abnormal **secretions**. Cells that pass the rigorous testing are kept growing in cultures or are frozen in a process called **cryopreservation**. Cryopreservation is a special freezing process in which cells or tissues can be kept frozen at temperatures less than at -70°C for months or years. Cells removed from cryopreservation must be tested when they are unfrozen for use in research and medical applications.

Medical Applications of Stem Cells

Bioprocessing involves the cultivation of multipotential and pluripotential cells, which are then induced to make a particular type of chemical.[6] Human embryonic stem cells are currently being used in bioprocessing to produce antibodies. A highly specialized technique called **expression engineering** can be used to create stem cells that produce therapeutic compounds such as hormones.

Recombinant stem cells have also been developed. These cells are genetically modified to produce a specific chemical having commercial or medical value. Genetic modification involves the addition of one or more genes so that the stem cells grow a specific gene product. Stem cells can be grown under conditions that stimulate the production of large amounts of material.

Figure 1.4 A research scientist prepares stem cell cultures in an incubator at Harvard University in Cambridge, Massachusetts. Due to ethical concerns, only a few states, including Massachusetts, actively encourage embryonic stem cell research.

Another approach uses stem cells to regenerate damaged or deficient body tissue. Stem cell technology has its greatest potential as a medical treatment in this area. Some critical areas of research are burn treatment and skin repair, **cardiovascular** restoration, **diabetes** treatment, **hematopoiesis**, immune system enhancement, and nerve cell regeneration. Contemporary medical practices often treat symptoms without curing the disease. It is hoped that stem cell practices will permanently restore normal levels of functioning. In addition, scientists hope to use stem cell regeneration as a means of replacing **transplantation** technologies. Transplantation is the replacement of tissue with tissue from the person's own body or from another person. However, it is not an ideal treatment, because the procedure often produces an immune response that damages or destroys the transplanted tissue.

The most common strategy for burn treatment and skin repair involves the transplantation of new skin. Donor skin is used in severe burns in which large amounts of skin are lost. Skin from the patient can be moved from one part of the body to another. The technique involves stretching existing skin to restore the damaged area. Since 1989, biotechnology companies have grown artificial skin in cultures. However, artificial skin lacks the nerve cells and many glands in normal skin functions. Only small areas of skin can be induced to grow back quickly by stimulating the damaged area with growth factors. In the future, scientists hope to use stem cells to regenerate large sheets of complete skin in a culture or on the damaged areas of skin.

Blood vessel and heart disorders, or cardiovascular diseases, are a major cause of death in Europe and North America. Cardiovascular diseases affect a large number of children, as well as millions of older adults. Diet, genetic factors, and such habits as smoking are three factors that damage blood vessels and the heart. A variety of surgical procedures are used to correct damaged and diseased blood vessels. Although procedures may improve the life of the patient, they do not completely repair the damage caused by the disease.

Researchers are currently working on cardiovascular restoration experiments using pluripotential stem cells. Recent studies on rats showed that hematopoietic stem cells will replace heart muscle cells damaged by the same conditions that cause a heart attack. Scientists have been able to use stem cells to correct muscle damage in other internal organs and skeletal muscles as well. Stem cell studies have been successful in repairing and replacing the lining of damaged blood vessels in rats. The stem cells used in these studies are differentiated in culture and then placed in the damaged structure where they regenerate the damaged structure. It is hoped that in the future hematopoietic stem cells can be used to diminish the effects of aging on the heart and blood vessels.

Hematopoiesis is the process the body uses to replace blood cells. Stem cells in the bone marrow differentiate into the various blood cells needed by the body. Blood is almost always in high demand by the medical community. It is essential to replace blood lost during major injuries and surgery. Blood is also important in cancer treatment. **Chemotherapy** treatments regularly kill off blood cells that need to be replaced to keep the patient healthy. The regular shortage of blood has compelled scientists to seek artificial blood substitutes. The current practice of transplanting hematopoietic tissue into a patient leads to medical complications such as **rejection** of the transplanted tissue.

Most of the substitute bloods do not carry out all the important jobs of actual blood. Researchers are currently using hematopoietic stem cells to grow particular types of blood cells and whole blood. It is hoped that in the future biotechnology and pharmaceutical companies will be able to produce thousands of liters of various types of blood to meet any need. Initial investigations indicate that blood can be produced on demand without the need for long-term storage. **Gene therapy** studies show that genetically modified hematopoietic stem cells can be implanted in people with blood disorders. The body can then produce normal blood for the rest of the patient's life.

Diabetes is a disease in which the body fails to control the levels of sugar in the blood. There is at present no cure for diabetes. Diabetes is typically treated with hormones or diet according to the underlying condition producing the disease. Severe cases of diabetes are usually due to an inability of the pancreas to produce the hormone **insulin**. Insulin helps the body's cells reduce the amount of glucose, or sugar, in the blood. High levels of glucose in the blood can damage the blood vessels, heart, and kidneys. Physicians typically treat diabetes by prescribing insulin treatments in the form of either injections or oral tablets. Insulin treatments are not fully effective at controlling blood sugar fluctuations. Currently, scientists are conducting studies that will allow them to implant stem cells in an effort to regulate insulin levels.

Immune system enhancement uses stem cells to enhance the body's defenses against autoimmune disorders, or **autoimmunity**, a condition in which the body's immune system attacks tissue it recognizes as foreign. **White blood cells** then produce antibodies against certain chemicals found in specific tissues. Diabetes, **lupus**, and **rheumatoid arthritis** are examples of autoimmune disorders.

Scientists are currently performing experiments using transplanted hematopoietic stem cells to destroy the autoimmune cells. In addition, they are replacing the patient's hematopoietic stem cells with new ones that do not attack the body. Another strategy involves the transplantation of genetically altered cells from the patient. The genetic alteration inhibits the white blood cells from attacking the body.

WHAT IS CLONING?

Cloning is an old technology derived from the natural ability of cells and certain organisms to replicate themselves. Scientists have conducted many experiments in which they were able to induce replication, or cloning, of cells and simple organisms grown in culture. The first cloning studies were conducted on plants as early as 1928. By the 1950s, special conditions had been developed that permitted the propagation of many crop plants using precise cloning techniques.[7] Plant cloning is done by removing cells or small parts of a plant using sterile cutting instruments. The plant materials are then transferred to a sterile culture medium containing nutrients. A growth of cells called a **callus** forms within a few days. Plant hormones are then added to the callus. A particular mixture of hormones causes the callus to form roots or stems. One to several young plants can grow from the callus. These plants are then transferred to soil where they grow into mature plants.

Animal cloning has proven a more difficult process to develop. One problem is that animal cells are hard to grow in culture. In addition, the growth conditions and hormones needed to induce development from cloned eggs are very complex. The first successful studies on cloning animals were performed on frogs in the 1962. John Gurdon of Oxford University cloned frogs from adult cells by placing the DNA of mature frogs into eggs. In 1979, Karl Illmensee, an embryologist at the University of Geneva, conducted experiments in which he cloned three mice using Gurdon's technique. Other researchers were not able to replicate Illmensee's experiments, leading to claims of fraud.

Figure 1.5 The genetically cloned sheep, Dolly, is photographed on February 25, 1997, at the Roslin Institute in Edinburgh, where she was created. Dolly was derived from adult mammary cells and is named after the famed singer/songwriter Dolly Parton. Sadly, Dolly the sheep died of lung cancer in 2003.

Another cloning strategy was developed in the early 1980s. A process called embryo splitting allowed scientists to produce several identical sheep and cattle by separating the cells of young embryos. Each cell removed from the original embryo was able to grow into a new identical embryo. Then, in 1986, Steen Willadsen of the British Agricultural Research Council cloned the first cattle from differentiated cells in a technique called reproductive cloning. Reproductive cloning uses the DNA of an adult animal to re-create that animal in the **fertilized egg**. The fertilized egg can come from that individual animal or a related animal.

Cloning made the international news in 1997 when Dolly the sheep was cloned at the Roslin Institute in Edinburgh, United Kingdom (Figure 1.5). Scientists created Dolly using a specific reproductive cloning technique called somatic cell nuclear transfer (Figure 1.6). **Somatic cells** are differentiated body cells. Dolly was created by removing the DNA from the **mammary gland** of one sheep and placing it in the egg of another breed of sheep. This egg was then placed in a female sheep called a surrogate. The egg then matured into a clone of the sheep that had donated the mammary gland DNA.

Unfortunately, Dolly died at 6 years old in 2003. Most sheep of its breed live approximately 12 years. Dolly died of lung cancer possibly induced by aging of the DNA that takes place in mature somatic cells. Scientists now know that the DNA of mature cells becomes damaged in several ways over the lifespan of an animal. This DNA damage can lead to cancer and the malfunction of cells. Other animals cloned using somatic cell nuclear transfer had similar fates. Scientists are now developing ways of correcting or reducing the DNA damage that occurred in Dolly. This opens the door for successful cloning trials that produce healthy adult animals. It is feasible to use this technology to clone humans. However, there are many ethical considerations and governmental regulations that restrict the cloning of humans. In addition, at this time there is no evidence that humans can be successfully cloned and grow to adulthood.[8]

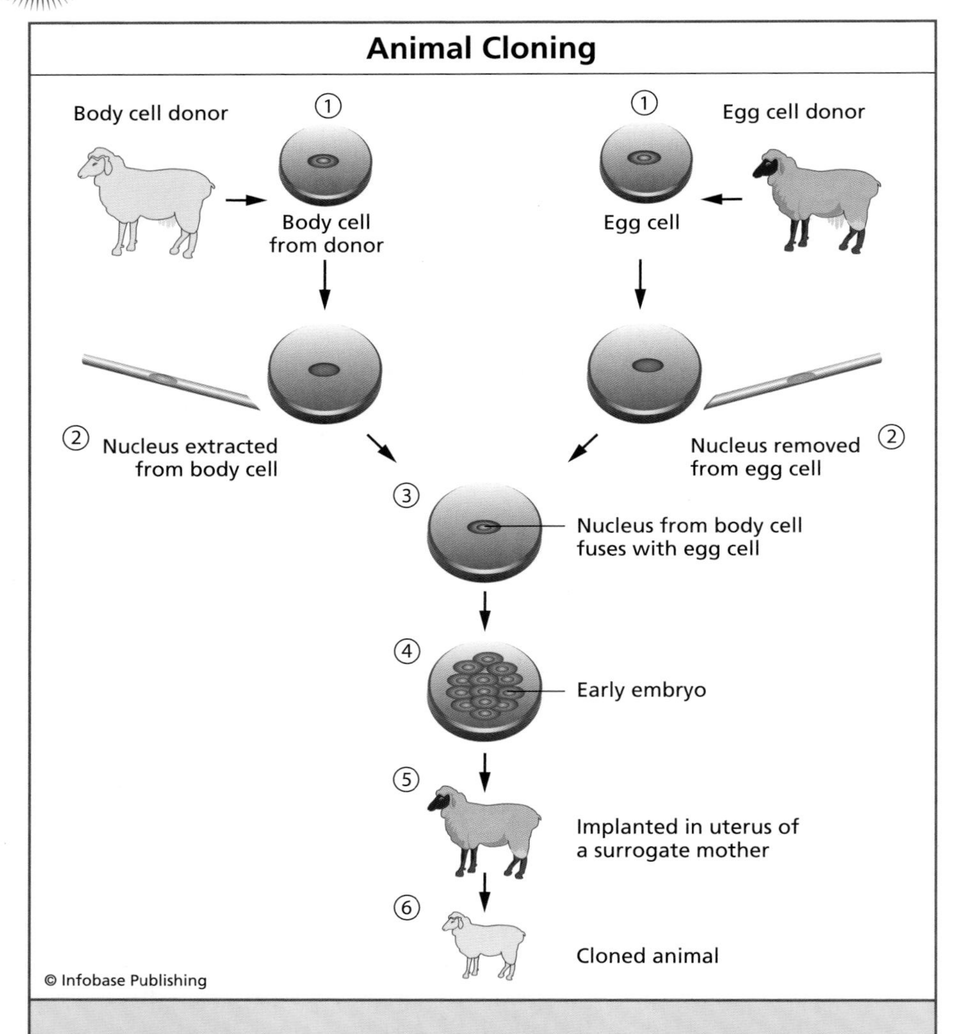

Figure 1.6 Scientists use somatic cell nuclear transfer as a means of cloning animals. Dolly the sheep was cloned using this technique.

Using both cloning and genetic engineering techniques, scientists can produce genetically modified (GM) clones.[9] Dozens of agricultural plants and several domesticated animals have been genetically modified and cloned for commercial purposes. One company uses a cloning technique developed by Texas A&M University to clone pet cats and dogs. Even a GM clone pet fish called the GloFish, a glow-in-the-dark zebra fish, is sold for home aquariums. The same techniques used to produce cloned animals are now being applied to medical therapies for abating human aging, injury, and disease.

CURRENT AND FUTURE USES OF CLONING

Medical uses of cloning are being directed into a group of innovative strategies called **therapeutic cloning**. Therapeutic cloning is the use of somatic cell nuclear transfer to produce stem cells that differentiate into tissues. Studies dating from the 1920s show that it is possible to differentiate an egg by replacing its nucleus with one removed from a mature cell. In effect, the egg can be differentiated into a liver cell by inserting the nucleus of a liver cell. The same strategy can be used on stem cells to grow clones of cells that then develop into a particular tissue or even a whole organ. Scientists are planning on using this strategy of therapeutic cloning to create replacement tissues and organs for transplants.

The most effective application of therapeutic cloning being developed today for human use involves the extraction of DNA from a patient needing a piece of tissue or a whole organ for transplantation. The patient's DNA is inserted into a donor egg that had the nucleus removed. The egg is then placed in a special culture medium where it divides into an early embryo. Cells from the embryo are harvested as though they were stem cells. This is achieved by growing the cells under special conditions that convert the cells into multipotential or pluripotential stem cell lineages. These artificial stem cells are placed into the damaged organ where they divide and differentiate to repair the organ. The body does not reject this new organ because its cells were all derived from the DNA of the patient.

A major limitation of this strategy is identifying the growth factors and cell culture conditions needed for the embryonic cells to differentiate into specific lineages of stem cells. In 2001, researchers at a biotechnology company called Advanced Cell Technology cloned the first human embryos raised in a culture. Only one embryo was successful at making cells suitable for differentiation into stem cells. This success led to the company's first trial with a cloned transplant. In 2002, the company transplanted cloned kidney tissues into cows by using a cloned cow embryo containing the DNA from the skin cell of a cow's ear. The knowledge gained from cloning Dolly made this potential therapeutic cloning technology possible. Scientists were also able to develop the embryos into a **fetus**, the later stage of development before birth. These fetuses were used as supplies of multipotential or pluripotential stem cells that could be transplanted and grown into differentiated adult organs. Immune system rejection does not take place during such a transplant because the cloned organ expresses the same antigens as the animal receiving the cloned organ.[10]

This line of research has raised fear of developing companies whose purpose is "fetus farming." Fetus farming is the use of fetuses as a living source of stem cells that can be harvested into transplant organs. This concern comes from the fact that there already is a global demand for organ harvesting from recently deceased people. However, these organs are only useful for individuals who are compatible with the organs. *Compatible* means that the person's body does not violently reject the organ. Because organs vary widely in the antigens that they express, there is very little likelihood that a harvested adult organ will match a particular person in need of the organ.

Parkinson's disease, which is a progressive degenerative brain disorder, has been the subject of many arguments supporting therapeutic cloning using fetuses. The disease is traditionally treated by administering a brain chemical called **dopamine** to remedy the shaking and uncontrolled muscle movement produced by the brain destruction. This treatment is not a cure. It must be given throughout a person's life to alleviate the symptoms of the disease. In the late 1980s, physicians experimented with transplantation as a treatment for Parkinson's disease. They transplanted a small piece of an adrenal gland into the brains of patients with Parkinson's disease. The adrenal gland, located just above the kidney, produces dopamine. Patients showed some alleviation from the disease. However, the piece of adrenal gland eventually died because the brain could not sustain it with adequate blood and nutrients.

More recently, researchers tested the efficacy of cloned fetal nerve cells secreting dopamine. It was hypothesized that fetal cells would reproduce within the brain and obtain the appropriate blood and nutrients needed to stay alive and function properly. A similar hypothesis is being tested for using this strategy to treat the death of brain cells associated with Alzheimer's disease. It is hoped that the success of the paralyzed rat experiment performed in 2006 at Washington University School of Medicine will support the success of this strategy of treating disease with cloned cells.

Another cloning strategy uses unipotential and multipotential cells from mature tissues to regenerate body parts. These cells have been isolated from a variety of organs and can be removed from living or recently deceased humans. Biotechnology and pharmaceutical companies worldwide have produced cloned artificial skin for burn victims. The skin has many of the protective properties of natural skin and may grow many of the features of natural skin over time. The success of artificial skin opens the door for cloning other adult organs. The major limitation with using cloned adult cells is a lack of knowledge about the growing conditions needed to direct the cells to regrow tissues and organs.

Many advances in cloning and stem cell research have been made since 1999, providing more promise for future use. Ethical issues related to using cloned cells and stem cells from embryos and fetuses confound the growth of this medical strategy. These concerns will continue to be addressed as scientists seek out acceptable practices for using cloning and stem cell applications.

REFERENCES

1. N. Seppa. "Stem Cells Repair Rat Spinal Cord Damage." *Science News*. 157 (2000): 6.
2. Stem Cell Research Foundation. "What's New." Available online. URL: http://www.stemcellresearchfoundation.org/WhatsNew/April_2006.html#4. 2006.
3. D.R. Marshak, R.L. Gardner, and D. Gottlieb, eds. *Stem Cell Biology*. Cold Spring Harbor, N.Y.: Cold Spring Harbor Laboratory Press, 2001.
4. S. Sell. *Stem Cells Handbook*. Totowa N.J.: Humana Press, 2003.
5. National Institutes of Health. "Stem Cell Information." Available online. URL: http://stemcells.nih.gov/info/basics/. 2006.
6. C.D. Helgason and C.L. Miller, eds. *Methods in Molecular Biology: Basic Cell Culture Protocols*. Totowa N.J.: Humana Press, 2004.
7. D.W.S. Wong. *The ABCs of Gene Cloning*. New York: Springer, 2006.
8. A. Kiessling and S.C. Anderson. *Human Embryonic Stem Cells: An Introduction to the Science and Therapeutic Potential*. Boston: Jones and Bartlett Publishers, 2003.
9. A. Chiu and M.S. Rao. *Human Embryonic Stem Cells*. Totowa N.J.: Humana Press, 2003.
10. National Research Council. *Scientific and Medical Aspects of Human Reproductive Cloning*. Washington, D.C.: National Academy Press, 2002.

FURTHER READING

Books

Alberts, B., J. Lewis, M. Raff, A. Johnson, and K. Roberts. *Molecular Biology of the Cell.* London: Taylor & Francis, Inc., 2002.

Alcamo, I.E. *DNA Technology: The Awesome Skill.* Philadelphia: Elsevier, 2000.

Bains, W. *Biotechnology: From A to Z.* New York: Oxford University Press, 1998.

Belval, B. *Critical Perspectives on Stem Cell Research.* New York: Rosen Publishing, 2006.

Borem, A., F.R. Santos, and D.E. Bowen. *Understanding Biotechnology.* San Francisco: Prentice Hall, 2003.

Fukuyama, F. *Our Posthuman Future: Consequences of the Biotechnology Revolution.* New York: Farrar, Straus and Giroux, 2002.

Herold, E. *Stem Cell Wars: Inside Stories from the Frontlines.* New York: Palgrave Macmillan, 2006.

Krimsky, S., and P. Shorett, eds. *Rights and Liberties in the Biotech Age: Why We Need a Genetic Bill of Rights.* Lanham, Md.: Rowman and Littlefield Publishers, 2005.

Thieman, W.J., and M.A. Palladamo. *Introduction to Biotechnology.* San Francisco: Benjamin Cummings, 2003.

Web Sites

Access Excellence

http://www.accessexcellence.org

Action Bioscience

http://www.actionbioscience.org

Biotechnology Institute

http://www.biotechinstitute.org

Cloning in Focus

http://learn.genetics.utah.edu/units/cloning/

Cold Spring Harbor Laboratory
http://www.cshl.org

Human Genome Project
http://www.ornl.gov/sci/techresources/Human_Genome/home.shtml

National Institutes of Health
http://stemcells.nih.gov

Science News Online
http://www.sciencenews.org

Gene Therapies and Tissue Engineering

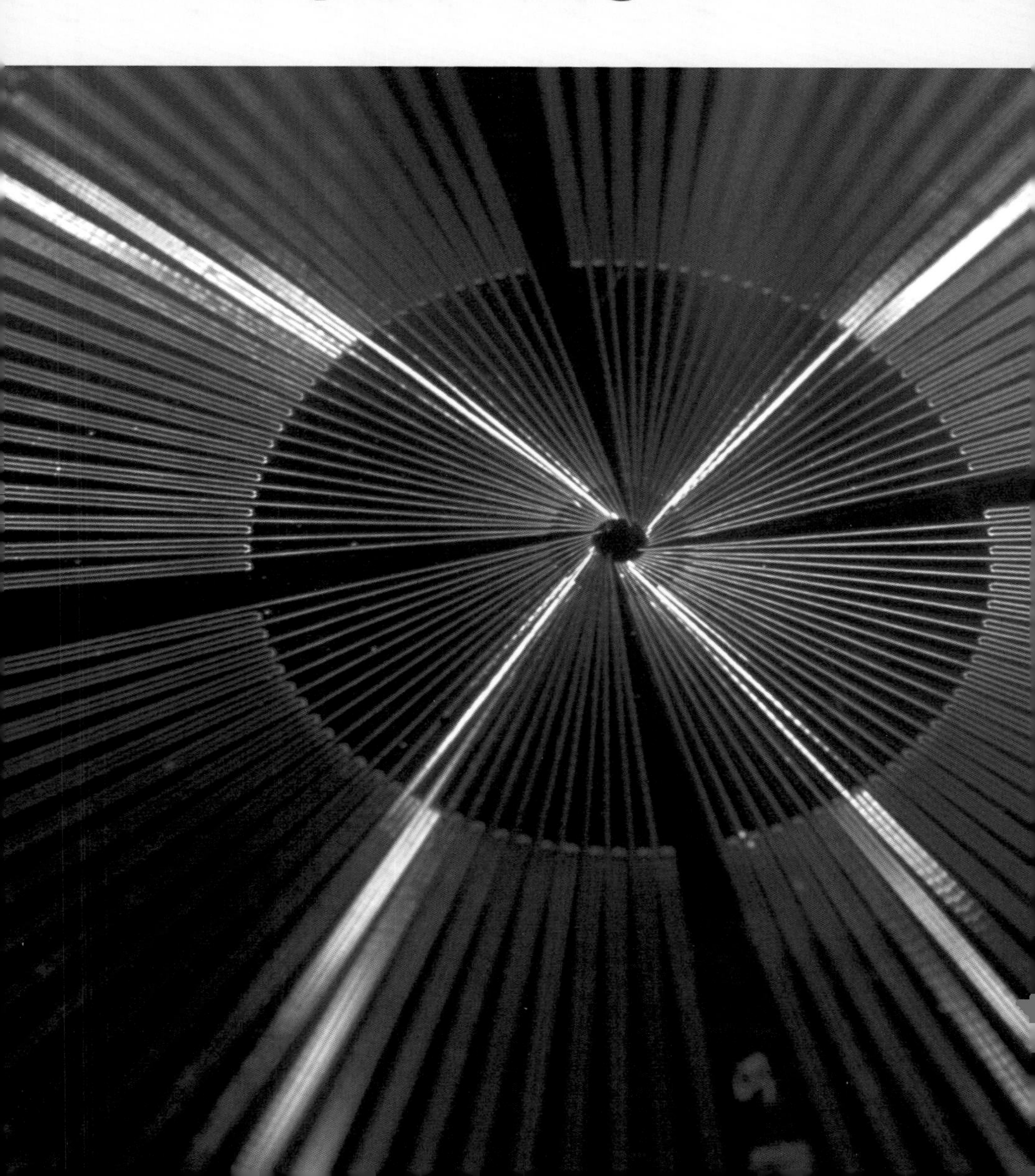

WHAT'S IN THE NEWS?

Ink-jet printers may be losing ground to laser printers, but they are becoming an innovative tool in molecular medicine laboratories. A team of researchers adapted a desktop printer for use in a machine that synthesizes cylinders of artificial living tissues. The technique was developed by Vladimir Mironov, a research physician at the Medical University of South Carolina, and Thomas Boland, a bioengineer at Clemson University.[1] Boland and Mironov call their procedure *organ printing* and describe it as a means of making three-dimensional living tissue (Figure 2.1). This three-dimensional technique is an improvement over other bioengineering technologies that produce flat prints of biological molecules.

The printing technique developed by Boland and Mironov uses cells that are poured into modified ink-jet cartridges. A computer using special software then sprays the suspension of cells to produce three-dimensional structures resembling tissues. Cells are placed in a special gel that hardens into a solid, custom-made shape. Boland sees his and Mironov's invention as a means of making transplant organs and tissues. He believes that one day modified printers will be used to make human skin for burn victims and other organs that are otherwise not available due to donor shortages. Initial trials are being done to produce artificial skin using sprayed protein structures containing skin cells.

In addition, Francesco Stellacci, a scientist in the Department of Materials Science and Engineering at the Massachusetts Institute of Technology, modified a printerlike machine to produce sheets of DNA called DNA-based chips. He called the technique nanoprinting. DNA nanoprinting produces single-stranded sheets of DNA that can be assembled in complex DNA patterns on microchips. These microchips, or **microarrays**, can be used as miniature tools for diagnosing genetic diseases in humans. Certain microchips can be developed to perform genetic identification of animals, humans, and plants. Nanoprinting can be adapted to make sheets of

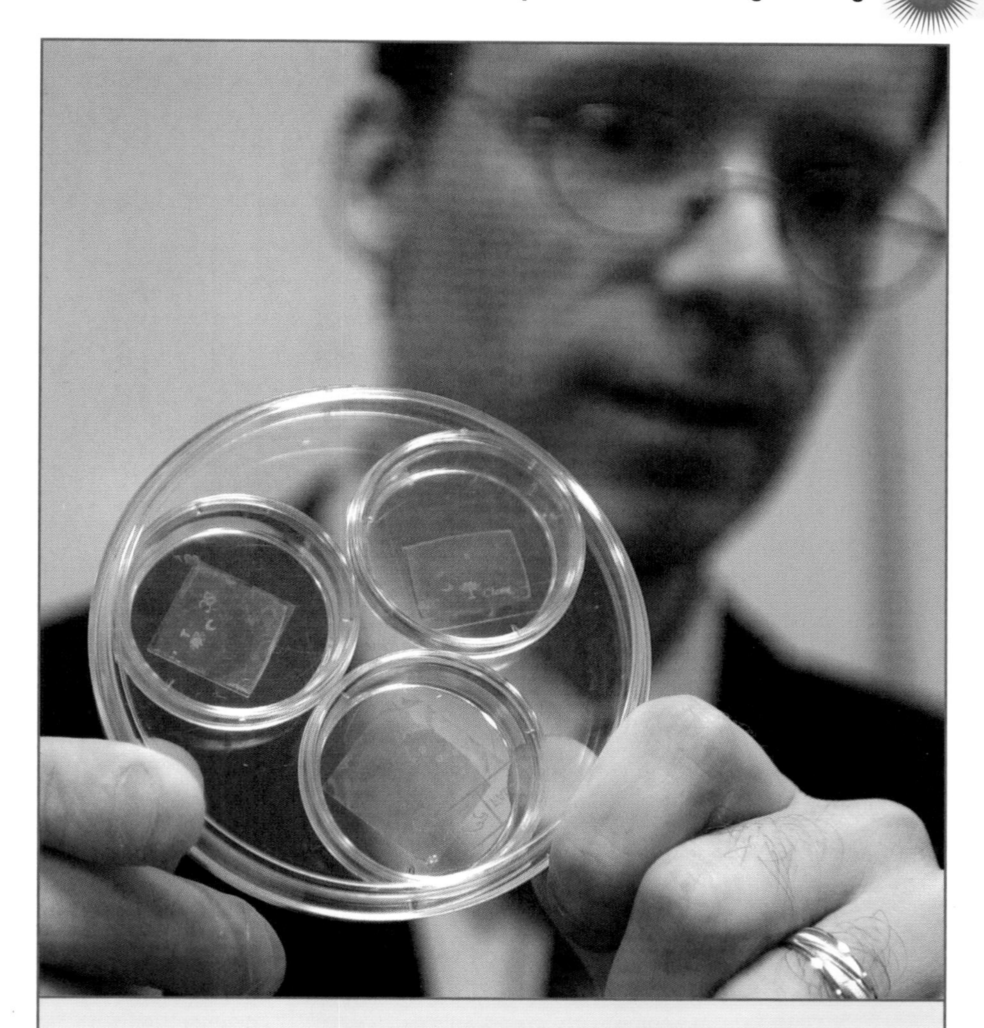

Figure 2.1 Bioengineer Thomas Boland and researcher Vladimir Mironov developed a biological engineering procedure called organ printing, which is used to manufacture three-dimensional living tissue. In the photograph above, Boland displays an example of his breakthrough technology.

carbohydrates that can be woven into artificial tissue scaffolds. Tissue scaffolds are used to make artificial tissues by implanting to the body or by adding cells grown in the lab.[2]

INTRODUCTION

Humans have been trying to combat disease for thousands of years. Medical practices developed in ancient Egypt were one of the first attempts at using scientific principles for treating disease and injury. Archaeologists discovered descriptions of healing practices written in **papyruses** dating from 3000 B.C. The most famous document is the Ebers Papyrus, which was primarily a medical reference describing a variety of diseases and injuries to the eyes, extremities, reproductive tract, and skin. More than 400 drug formulations and many surgical procedures are described in the papyrus. Many of these treatments resemble modern medical practices. For example, aspirin (salicylic acid) and morphine were commonly used to treat aches and pains.

Just like the ancient Egyptians, modern physicians recognize that certain ailments and injuries are difficult to treat. Genetic disorders make up the largest group of disease that has no simple cure. Any condition caused partly or completely by a defect in one or more genes is called a genetic disorder. Most genetic disorders lack therapies because curing a genetic disorder would involve modifying the DNA of the cells in the defective gene. Cancers also eluded treatment by ancient medical practices. Even today, traditional cancer treatments are not always effective and can be very harmful to the patient.

Humans have been attempting to treat injuries and the effects of aging throughout history. The early Greek and Roman civilizations used surgical repair and simple **prosthetic devices** to repair tissues and organs damaged by war and work-related injuries. Prosthetic devices are objects or machines that replace damaged or missing body parts. Amputation, or the surgical removal of a limb, was commonly done when the damage exceeded the ability for the body to heal. The missing body part was then replaced with a prosthetic device. Ancient prosthetic devices were simple crutches or leather cups. Some of these devices afforded a wide range of motion. The Roman general Marcus Sergius was outfitted with an iron hand after he lost his hand during the Second Punic War.

Modern prosthetic devices were developed to offset lost body functions resulting from amputation and body part damage. However, these mechanical devices did not possess the same function as the original human tissue. The advent of modern biotechnology in the early 1970s gave us new ways of replacing lost body functions. Gene therapy and tissue engineering are two promising areas of biotechnology intended to cure genetic disease or to regenerate human tissue.

WHAT IS GENE THERAPY?

The American Association for the Advancement of Science defines gene therapy as the altering of genes in order to affect their function. A definition provided by the Human Genome Organisation (HUGO) has a more detailed description of gene therapy. They describe it as an experimental procedure aimed at replacing, manipulating, or supplementing nonfunctional or misfunctioning genes with healthy genes.

Gene therapy is based on the principles of the first successful genetic engineering trial. In 1973, Herbert Boyer, of the University of California at San Francisco, and Stanley Cohen, of Stanford University, collaborated to create the first **recombinant DNA** organism.[3] Recombinant DNA is a novel sequence of genetic material that is formed by combining pieces of DNA from different organisms or cells. Boyer and Cohen produced the organism using genetic manipulation techniques developed by Paul Berg in 1972 at Stanford University. Berg's epic research involved splicing new genes on a small circular piece of bacterial DNA called a **plasmid**. He used chemical tools called **enzymes** to cut and paste the DNA segments.

The special enzymes used in Berg's research laid the foundation of recombinant DNA technology. Berg used two categories of enzymes to make the plasmid: **restriction enzymes** and **ligase**. Restriction enzymes were discovered in 1970 by Hamilton Smith and Kent Wilcox at Johns Hopkins University. Restriction enzymes are bacterial enzymes that cut DNA at specific locations determined by a certain chemical sequence of DNA. Ligase is an enzyme used to combine the cut ends of DNA fragments. DNA ligases were first identified and purified in five separate laboratories in 1967, when researchers came across the enzyme while investigating DNA replication.

Cohen and Boyer were the first people to perform Berg's technique on living bacteria. The two scientists used the cutting and splicing technique to create bacteria with two different antibiotic resistance genes. The research of Cohen and Boyer heralded the development of many biotechnology applications used to make a variety of commercial products and medicines. However, it did not confirm the likelihood of using genetic engineering on complex organisms such as humans.

Gene therapy became plausible following a series of investigations on organisms having a cellular structure and a DNA complexity similar to that of humans. In 1977, biochemists Bill Rutter and Howard Goodman at the University of California at San Francisco isolated the gene for rat insulin. Insulin is a hormone that helps adjust blood sugar. The hormone is deficient in people with type 1 diabetes. This discovery was followed by two projects conducted at Genentech, which is a biotechnology company in California. In 1977, the company was able to express a molecule called **somatostatin** in bacteria. Somatostatin is a protein that stimulates the production of human growth hormone. Then, in 1978, Genentech and the City of Hope national medical center produced human insulin in bacteria using recombinant DNA technology. Science now had available the tools to successfully transfer working copies of human genes to other organisms. Another biotechnology company called Chiron Corporation then developed the tools to express human genes in yeast cells in 1981. Yeast cells are very similar in function and structure to human cells. This discovery confirmed that genetic modification was possible in human cells.[4]

The final step leading to human gene therapy came about in 1985 with the development of the "Harvard mouse" (Figure 2.2). It

Figure 2.2 Dr. Philip Leder, a molecular geneticist at Harvard University, holds a transgenic mouse known as the Harvard mouse. The transgenic mouse was genetically modified to carry a cancer-promoting gene called an oncogene.

was developed by Philip Leder of Harvard University and Timothy Stewart of the University of California at San Francisco. The Harvard mouse was the first **transgenic** mouse and it fueled a controversy about getting a **patent** for living organisms. *Transgenic* refers to an organism that has genes from another organism inserted into its DNA using recombinant DNA techniques. A patent is a government grant giving an inventor the exclusive right to make or sell an invention for a term of years. The Harvard mouse, also known as the Oncomouse, was genetically altered to have a human cancer-promoting gene called an **oncogene**. It was created by injecting the oncogene into fertilized mouse eggs. The eggs were then implanted into a female mouse, which then gave birth to the transgenic offspring. All the cells of the offspring mice contained the oncogene and were able to pass the gene on

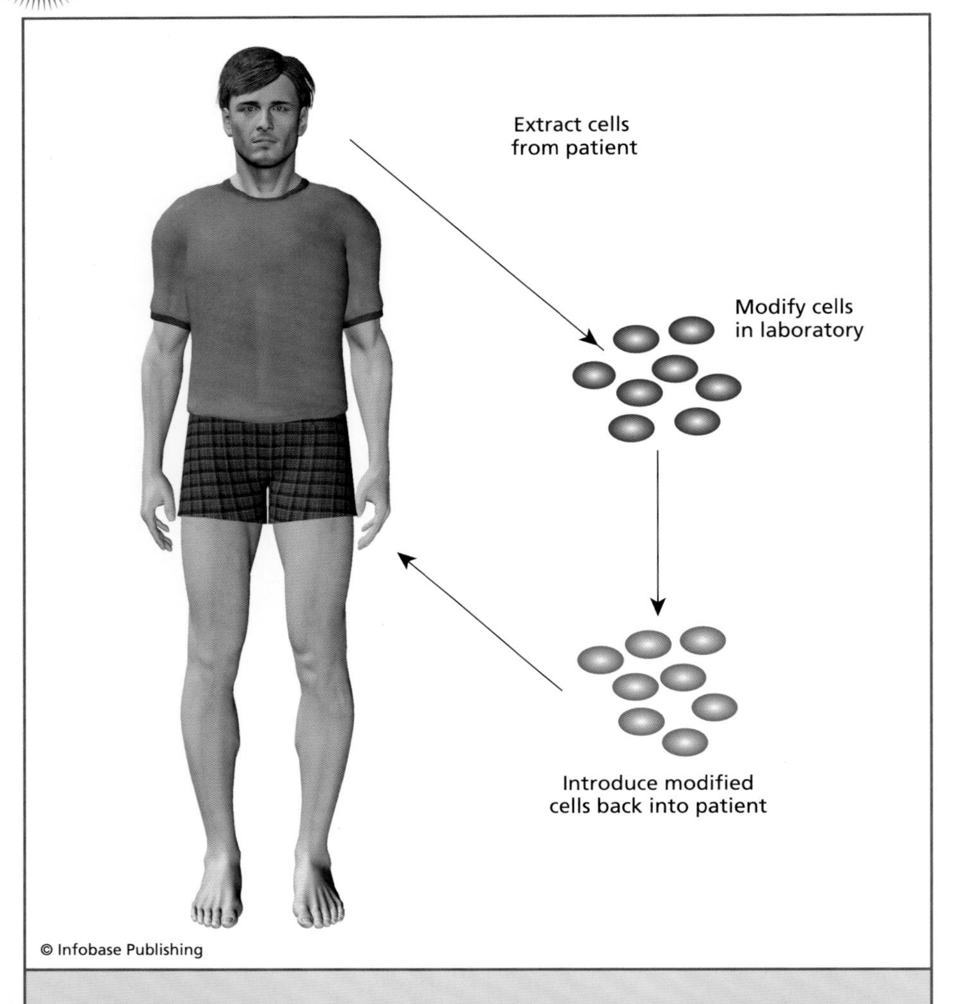

Figure 2.3 Human cells can be removed from the body, genetically modified in a laboratory, and returned to the body to treat genetic diseases.

to future generations. The Harvard mouse was more likely to produce cancer than were other mice and was useful for human cancer studies.

The Harvard mouse paved the way for the first gene therapy trial in humans, performed in 1990. A team of researchers at the United States National Institutes of Health in Bethesda, Maryland, performed the first approved gene therapy on a four-year-old girl named Ashanti DeSilva. The team, led by W. French Anderson, Michael Blaese, and Kenneth Culver, corrected a rare genetic disease called **severe combined immune deficiency** (SCID). SCID disables the immune system, making it impossible for the body to fight off disease. They did this by removing the girl's white blood cells and genetically modifying the cells to have a healthy form of the SCID gene (Figure 2.3). The genetically modified white blood cells were then placed back into the girl's body. She was then able to fight off disease as if she did not have SCID. This first human trial on gene therapy opened the way for gene therapy to cure a wide array of disorders.

CURRENT AND FUTURE USES OF GENE THERAPY

Gene therapy has progressed to a great extent since the first trial conducted by Anderson and his team in 1990. Ashanti DeSilva has lived longer than any other person born with her life-threatening condition. She still needs regular treatments of genetically modified cells. Nevertheless, she would have died many years ago from a variety of infections if only traditional treatments had been used to fight her illness. The scientific community has had many successes and failures with gene therapy since 1990. However, the promises of gene therapy were seriously questioned in 1999 with the death of 18-year-old gene therapy patient Jesse Gelsinger. The death caused a team of scientists at the University of Pennsylvania to discontinue their gene therapy experiments for treating a rare liver disorder called ornithine transcarbamylase (OTC) disorder. People with the disease cannot remove ammonia, a dangerous waste product, from the blood.[5]

Gelsinger's gene therapy treatment was very different from the type performed on Ashanti DeSilva. The treatment done on Gelsinger produced a severe immune system reaction that ultimately caused his death. Government investigations of the research showed that some safety protocols were not followed during Gelsinger's therapy. This setback was followed by the deaths of two French boys in 2003 that were successfully treated for SCID in 2000. It was discovered that the gene therapy technique used at Necker University Hospital in Paris to treat the SCID turned on genes that produced **leukemia**. Leukemia is a cancer of the bone marrow that produces white blood cells. Other children given the same treatment at Great Ormond Street Hospital in London using the same technique did not develop leukemia. However, petitions by various patient advocates and other groups stirred public controversy about gene therapy.

As a result of these circumstances, gene therapy treatments are considered experimental and are not approved for any clinical use by the United States Food and Drug Administration (FDA). The FDA is a governmental agency that oversees the effectiveness and safety of medicines and medical treatments. Government approval for clinical use is necessary for a therapy to be an accepted medical practice. Many gene therapy researchers have followed a voluntary ban on certain types of gene therapy until safety concerns are fully addressed. The progress of gene therapy has slowed to a cautious pace. However, new treatments are still being conducted on animal models, and careful human trials are still being conducted throughout the world.

Ex Vivo Technique

Anderson's team used a clever technique to treat DeSilva's condition. This classical and valuable gene therapy technique is called "ex vivo somatic cell gene therapy."[6] **Ex vivo** means the procedure was conducted in an artificial environment outside the body. Somatic cells are body cells such as those found in the blood, liver, or lungs. The cells removed from Ashanti DeSilva's body were altered inside a laboratory (ex vivo) and then placed back into her body. Her blood cells, which are somatic cells (and explains why the technique is called somatic cell gene therapy), were targeted for the gene therapy because they expressed the disease trait. There was little need to alter other body cells to treat Ashanti's condition immediately. Ex vivo somatic cell gene therapy is restricted to genetic disorders that affect a specific grouping of cells that can be easily removed and placed back in the body.

This successful technique led other researchers to produce equally astounding results. In 1994, University of Michigan Medical Center scientists were able to treat a blood fat condition by genetically modifying a patient's liver cells. Researchers at Aston University in Birmingham, England, were successful using the technique to treat type 1 diabetes. They did a series of preliminary studies in 1999 in which they genetically modified a variety of cells to produce insulin in the body. Another direction for ex vivo somatic cell gene therapy is the correction of diseases caused by growth hormone deficiencies.

In Vivo Technique

Yet another type of gene therapy has resulted in even more promising treatments. **In vivo** somatic cell gene therapy uses **viruses** as a means of inserting correct forms of a gene into particular cells affected by a genetic disorder. *In vivo* refers to a procedure carried out within the body. As a result of this strategy, the cells do not have to be removed from the body, the way they did when doctors treated DeSilva's malady. Cells are genetically modified within the body. The viruses are designed to attach to a particular cell called a target cell. Once attached to the target cell, the virus enters the cell and deposits its genetic material. The viral genetic material is modified in the laboratory so that the virus inserts a desired gene while not causing disease in the target cell. An early study carried out in 1994 at Vanderbilt University used viruses with a gene that covers up a defective gene that causes lung cancer. The researchers used a **retrovirus** to insert the gene into lung cells. A retrovirus is a type of virus that has **RNA** instead of DNA as its genetic material.

Retroviruses were also used in studies at the University of California at San Francisco, Harvard University, and the National Cancer Institute in 1999 to treat a form of brain cancer in a mouse model. Similar treatments were effective in human studies and proved safer than traditional surgical and chemical methods of treating brain cancer. In 1997, scientists at the University of Florida used a different type of virus to treat genetic disorders. They used a virus called an **adenovirus** to insert normal genes that covered up the genes for cystic fibrosis in lung cells. Adenoviruses are a group of viruses that cause respiratory tract infections. In 2004, researchers at the University of Texas Southwestern Medical Center at Dallas successfully tested in mice the first gene therapy for skin cancer. They used an adenovirus carrying a normal gene to reduce cancer produced by a genetic disease called **xeroderma pigmentosum**. Xeroderma pigmentosum is an inherited disorder that causes extreme sensitivity to sunlight and early onset of skin cancers.

Germ Cell Therapy

Germ cell gene therapy studies on humans are currently voluntarily banned in the United States by the scientific community. *Germ cell* refers to the egg- or sperm-producing cells that are found in the gonads. Many scientists support the ban because all the biological consequences of genetically modifying germ cells are unknown. Promising findings are coming from studies performed on mice and monkeys. Since the mid-1980s, scientists have been using ex vivo and in vivo methods to insert human genes into animal eggs, embryos, and sperm. A host of these studies done in animals and funded by the United States National Institute of Child Health and Human Development has provided likely models of how germ cell gene therapy can be used to treat and prevent the inheritance of many common genetic disorders, including **cystic fibrosis**, **hemophilia**, **Huntington's disease**, and **sickle cell anemia**.

Cystic fibrosis (found in 1 of every 3,200 Caucasian births and carried by nearly 4% of all Caucasians) is a lethal genetic disorder that affects many cellular secretions, particularly those of the lungs. Most patients live beyond adolescence now with more effective life-prolonging treatments. Hemophilia is characterized by uncontrolled bleeding. Huntington's disease causes the loss of nerve cells in a specific part of the brain. Sickle cell anemia is a disease of the red blood cells that limits the blood's ability to carry oxygen. Gene therapy studies are also being performed on stem cells to see if genetically modified stem cells can grow into normal tissues and organs in the body.

WHAT IS TISSUE ENGINEERING?

Tissue engineering is an interdisciplinary science that brings together biologists, chemists, engineers, bioengineers, and physicians. The goal of tissue engineering is to use a replica of a living tissue to replace damaged or diseased tissues and organs. Biologists play a role in defining the functions and needs of the body parts being replicated. Knowledge of chemistry is essential for understanding the properties of the materials replacing body functions. Engineers are responsible for various design features that permit proper functioning of the replicated tissue while preventing premature wear and tear. Physicians are needed to implant and monitor the engineered tissues.[7]

Tissue engineering, an outgrowth of bioengineering, includes **biomaterials**, cell transplantation, controlled-drug delivery systems, and polymer chemistry. At its inception, bioengineering focused on diagnostic devices, medical devices, **prostheses**, and **medical imaging** equipment. Diagnostic devices are used to detect the presence of disease and include instruments used in a laboratory and devices implanted in the body. Medical devices are a variety of machines ranging from pacemakers to synthetic heart valves. Prostheses include items such as artificial joints that assist mobility and replace large body functions. Medical imaging uses a variety of tools to visualize body components using magnetic fields, sound waves, and X-rays.

Tissue engineering includes two specialized fields of study: tissue repair technology and tissue regeneration technology. Tissue repair technology, as the name implies, uses special materials and devices to replicate and replace the function of a tissue. "Scaffold-to-seed" technology is the primary method used in tissue engineering. It uses cells and substances called biomaterials to build artificial tissues that replace damaged or lost body components. Biomaterials are natural or synthetic substances that are suitable as a substitute for living tissue, especially as part of a medical device. Synthetic biomaterials are made from carbohydrates, **hydrocarbon polymers**, proteins, and silicone. Certain natural biomaterials are derived from treated bone, cartilage, coral, skin, and tendon. The materials must be modified so that they do not chemically damage or poison nearby tissues. In addition, biomaterials must not cause the body to produce an immune response that leads to rejection of the device made from the material. A term called **biocompatibility** is used to describe biomaterials that do not react with the body in harmful ways. The United States Food and Drug Administration has strict guidelines for determining the biocompatibility of biomaterials.[8]

Biomaterials are usually molded or woven into particular shapes that permit them to carry out a specific job in the body. They can be used to encapsulate or enclose metal or plastic mechanical devices with the goal of preventing the body from rejecting the implanted material. Some biomaterials are embedded with electrical diagnostic devices that monitor the body. Others have circuitry for operating the device or sending an electrical impulse through the biomaterial to make it carry out a function. It is possible to build a replica of a heart or of a gland using biomaterials incorporated into small machines.

In addition, tissue regeneration technology uses cells and biomaterials to facilitate the growth of damaged tissue. Tissue regeneration differs from tissue engineering because the goal is for the body to replace or regrow the damaged tissue.

Tissues are manufactured in three main steps either in a laboratory, **ex situ**, or in the body, **in situ**. Ex situ tissue regeneration starts out by growing cells in a culture that encourages the cells to divide (Figure 2.4). Healthy cells are then placed on a biomaterial kept under conditions that encourage the cells to replicate. The biomaterial structure is a called a **matrix** or **scaffold**. This matrix can be made of molded natural or synthetic substances. Usually, the biomaterial is placed in a container with a liquid that provides the cells with nutrients and growth factors that stimulate replication. This container holding the biomaterial is placed

Figure 2.4 A scientist dips a biodegradable mold, shaped like a bladder and seeded with human bladder cells, into a growth solution. The procedure, known as ex situ tissue regeneration, creates tissues that can be transplanted into a human patient.

in an incubator that mimics the growing conditions of the body where it will be inserted. Differentiation factors are then added while the matrix is kept in growing conditions that encourage the cells to act like the tissue that is being repaired.

If the synthetic tissue is functioning properly, it is transplanted into a patient. Regular testing is done to see if the body accepts the new tissue. Blood vessels ultimately grow into the matrix. Placing chemicals called **angiogenesis factors** into the matrix can sometimes stimulate the blood vessels to grow. Angiogenesis factors are chemicals that stimulate the development and growth of blood vessels. Eventually, cells from the body grow into the matrix replacing the cells that were grown in the laboratory. Sometimes the matrix is destroyed by the body and replaced by a natural matrix secreted by the body cells.

The first step of in situ tissue regeneration is the molding of a biomaterial matrix that fits into a damaged or lost component of the body. The matrix is coated with chemicals called growth factors to encourage cells from the body to grow. Differentiation factors can be used to facilitate the growth of the stem cells into various cell types. This technique only works if stem cells for the tissue being repaired are located near the matrix.

CURRENT AND FUTURE USES OF TISSUE ENGINEERING

Tissue engineering is a new science. The first paper recognizing tissue engineering as a valid field of medicine was a report titled "The Emergence of Tissue Engineering as a Research Field" prepared in 2003 by the National Science Foundation. The report summarized the importance of tissue engineering in the following statement:

> Consider, for example, the basic dilemma faced by a surgeon. While the removal of organs or body structures that are damaged beyond repair by disease or trauma can be life-saving, the patient must cope with the functional effects of tissue loss and, in some cases, the psychological impacts of disfigurement. And for those vital organs whose complete removal is incompatible with life, the surgeon's hard-earned skill is, by itself, of no avail: the procedure cannot be done unless there is some way of replacing or reconstituting essential functions.

Most of the applications of tissue engineering targeted in the report were for medical purposes. However, researchers are now discovering that artificially grown tissues have many potential uses beyond the medical applications.[9]

A symposium on tissue engineering, then called regenerative or reparative medicine, held at the University of California at Los Angeles in 1992 identified directions for contemporary tissue engineering. Since then areas that have evolved with the growth of tissue engineering research include:

- Cellular prostheses for the replacement of human body components.
- **Acellular** replacement parts that induce tissue regeneration.
- Tissue-like and organ-like modeling systems for applied and basic research.
- Cell and cell product delivery systems.
- Clinical or environmental monitoring systems using scaffolded cells.
- Biological surfacing on nonbiological devices.
- Computational devices using cells as neural networks.
- **Nanobiotechnology** devices using components of cells.

The number of published research projects involving tissue engineering in these categories increased from one paper in 1984 to 126 papers in 2001. During that period 685 scientific investigations were published. There was a decline in research during 2001 following legislation enacted by President George W. Bush that limited the use of human stem cells in the United States. In 2000, there were 214 investigations compared to the 126 studies in 2001. Approximately 65% of the studies conducted between 1984 and 2001 were basic research and were not used in patients at the time of publication. Many of the studies targeted the regeneration of blood vessels, bone, cartilage, and heart tissue.[10]

Cellular Prosthesis

Major breakthroughs are being made in the development of cellular prostheses for the replacement of human body components. Probably the most amazing direction of cellular prosthesis is for the replacement of sight in people with damage to the **retina**. The retina is the inner layer of the eye containing nerve cells that detect light. Harvey Fishman, M.D., Ph.D., Daniel Palanker Ph.D.,

Michael Marmor M.D., and Mark Blumenkranz, M.D. of the Department of Ophthalmology at Stanford University discovered that within 48 to 73 hours retinal cells will move into a perforated matrix. The cells implant in the matrix while keeping their connections to other nerve cells in the retina. The researchers were able to attach electrical circuits to the cells and the matrix. These electrodes are organized in a manner that mimics the way the eye sends information to the brain. This technology replaces earlier attempts at producing artificial eyes that use camera-like devices attached to the brain. These cameras are only able to produce shadowlike images for the brain. In a 2006 press release from Stanford University, Dr. Fishman expressed hope that the retinal prosthesis will provide the brain with the same images as those given by a natural eye.

Dr. Ravi Bellamkonda at Case Western Reserve University is currently investigating neural prosthesis to repair brain and spinal cord damage. His research team reported at the 2004 National Institutes of Health Neural Interfaces Workshop a way of using special cells called **glial cells** to regenerate a damaged brain and spinal cord. Glial cells mold and maintain the nervous system. Dr. Bellamkonda was able to use tissue engineering to produce a gel matrix that can be shaped to a particular region of the brain and spinal cord. Glial cells are then added to the matrix and they encourage the growth and organization of nerve cells placed in the matrix. The matrix can then be placed in the body where it connects to appropriate groupings of nerve cells to repair the damaged or lost brain or spinal cord tissue.

Several research teams in Europe and the United States are using a technique called **endothelial cell** seeding to produce artificial blood vessels. Artificial blood vessels are called vascular prostheses. These blood vessels are produced by first molding a matrix into the shape and size of the desired blood vessel. Cells called **fibroblasts** are then coated on the matrix. Fibroblasts grow into connective tissue that supports the blood vessel and may ultimately replace the matrix. The matrix is then seeded with endothelial cells. Endothelial cells form linings on organs and body cavities. They also form the lining of blood vessels. Older styles of prosthetic blood vessels were not made of biological materials and did not act like real blood vessels in the body. These older styles sometimes tore away from the blood vessels and formed dangerous clots that could block the vessel or break off and lodge in body organs.

Other body organs and structures are being targeted for cellular prosthesis. In 2005, researchers at King's College in the United Kingdom used a specialized seeded matrix to produce a replacement tooth that regenerates into the tooth socket of the jaw. Deafness may be a thing of the past because researchers at the University of Iowa are developing a way to make an artificial **cochlear**. The cochlear converts vibrations into sound and then transmits the information to the brain. A variety of studies is being conducted to replace or regenerate hormone-secreting organs such as the pancreas. Even the immune system has been the target of cellular prosthesis. Researchers have produced artificial **lymph nodes** using white blood cells and special beads that can be placed near an organ. Lymph nodes are immune system structures that help the body fight disease and that assist with the repair of tissues.

Acellular Replacement Prosthesis

Acellular replacement parts that cause tissue regeneration are primarily successful for repairing body parts containing bone, cartilage, ligament, and tendon. This is currently the most researched area of tissue engineering and shows the most promise for present-day applications. The hallmark study reported in the July 17, 1998, *Business Week* magazine took place at the University of Massachusetts. In this study, Dr. Charles Vacanti led a research team that used tissue engineering to regenerate bone found at the end of the thumb in a 36-year-old man who had crushed his thumb in an industrial machine. The team was able to create an artificial bone scaffold using the skeleton of a coral. Eventually, the coral disintegrated as it was replaced by the patient's bone cells, which regenerated a new bone. The same research team became famous for scaffolding a human ear on the back of a mouse (Figure 2.5).

Figure 2.5 The regeneration of body parts with acellular replacement parts is proving useful in the replacement of bone and cartilage structures. The mouse shown in the above image received the matrix for the cartilage that forms the human ear.

Skin regeneration relies on advances in acellular replacement technology. The traditional way of repairing skin requires removal and relocation of skin from the patient. This technique is only successful if the patient has enough healthy skin left to do the relocation. People with severe burns or bacterial infections of the skin do not always have much remaining skin. Newer strategies developed in the 1990s grow sheets of artificial skin. However, this skin lacks essential components such as hair, nerve cells, and sweat glands. Acellular matrices are being developed that encourage remaining skin cells to migrate onto the matrix. The new skin will contain all the components of normal skin.

Cancer Treatments with Artificial Tissues

Cancer research is one of the first sciences being improved using tissue-like and organ-like modeling systems. Artificial tissues and organs are being developed to better understand the ways cells in particular body regions become cancerous. Scientists are able to use the artificial tissues to investigate the chemical and environmental factors that make normal cells turn into cancer cells under body conditions. Current cancer research involved cells grown alone in cultures that do not fully represent the complexity of the body. Tissue-like and organ-like modeling systems are also providing a better understanding of human development. Scientists are able to use these model systems to investigate ways certain stem cells are able to create a particular organ. This is very important information that has many applications in regenerative medicine.

Living Machines from Cells

Cell and cell product delivery systems are "living machines" that replicate cells or produce cell secretions. These cells or secretions can then be collected and used for a variety of purposes. Japanese researchers at Kyushu and Osaka universities are working with the American Association for Cancer Research to make cell delivery systems that replace certain pancreas functions. They are hoping to use the technology to make "living insulin injectors" for diabetics. These delivery systems can be used as temporary artificial organs used to treat a variety of diseases. They can also be used to make a diversity of cell chemicals having many applications in medicine. Certain researchers are hoping to use these delivery systems to remove hazardous chemicals from water.

Cell delivery systems must be able to monitor the body in order to work properly. They require biosensors to detect particular conditions of the body. *Biosensor* is the traditional name for clinical or environmental monitoring systems that use cell components. The older technology used components of the cell membrane attached to electrical circuits that amplify and convert electrical signals into a readout. Cell membranes were recognized as being highly selective chemical detectors. Certain cell membrane proteins, called **receptors**, are able to bind to specific types of molecules. Upon doing so, they send a chemical or electric signal to the cell indicating the attaching of the molecule. These devices have been used to detect the presence of chemicals in air, blood, body fluids, drinking water, and foods. However, they are fragile and sometimes last for only one use. Scientists are developing similar devices using scaffolded cells. In effect, the cells can be made into a living machine that regenerates itself if damaged or worn out. Cellular monitoring systems can be incorporated into the body to monitor various functions throughout the lifetime of a person. In addition, they can be placed in a body of water to detect particular pollutants. Information from the monitoring systems can be transmitted using radio waves or cell phone communication technologies.

Biological surfacing on nonbiological devices conjures images of **cybernetic organisms**, or **cyborgs**, such as the main character played by Arnold Schwarzenegger in the 1984 movie *The Terminator.* The robot in *The Terminator* was covered with human skin in an attempt to cover up his identity as a robot designed to kill people. It is not expected that biological surfacing on nonbiological device technologies will lead to dangerous robots. Earlier attempts of covering nonbiological surfaces involve placing a layer of DNA or proteins on glass or plastic slides. These objects, called **biochips**, are usually used to detect the presence of particular types of biological molecules. Researchers at Lawrence Berkeley National Laboratory and the University of California at Berkeley are developing biochips that use whole cells. The scientists are hoping to assemble the cells into different types of tissues that can be used as diagnostic instruments. In addition, they can use the biochip as a way to study the effects of DNA damage to particular tissues. Other researchers are hoping to produce microscopic cyborgs for use in water treatment, food industries, and environmental monitoring.

Artificial Neural Networks

Advances in computational devices using cells as neural networks rely on neural prosthesis and biochip technologies. Artificial Neural Network, or ANN, is an information-processing technology that works with the same complexity as the brain or nervous system components. The first ANN was developed in 1969 and used traditional circuitry to carry out complex mechanical operations. It was based on a circuit called the artificial neuron developed by neurophysiologist Warren McCulloch and mathematical theorist Walter Pits in 1943. Computer software technology permitted ANN to develop decision-making capabilities that learned from experience and trial-and-error activities. Unfortunately, computers could not be designed to make decisions with the rapidity and accuracy of the human brain. Some researchers used this technology to make robot rats that learned how to run a maze. The robots even learned how to cheat at the maze tasks.

Scientists are developing ways of making ANN from nerve cells grown on matrices containing electrical circuits. Neural networks that use nerve cells can take complicated approaches to problem solving, and they develop reasoning patterns better than those of conventional computers. Conventional computers follow simple sets of instructions when solving a problem. Neural networks can create new reasoning patterns and therefore produce new knowledge. A cell-based neural network uses a neuron or pathway of neurons attached to an electric circuit. The circuit has a **transducer** that receives signals from particular regions of the neuron or pathway. A transducer is any device that converts one form of energy into another form of energy. It is hoped that neuron-driven neural networks could be used to operate software and hardware for diagnosing disease, running robots that drive cars or fly airplanes, perform surgery, interpret large volumes of complex data, and predict weather.

Nanobiotechnology Strategies

Most nanobiotechnology devices being developed today use biological molecules and other chemicals to make microscopic instruments or machines. However, cells can also be used to run nanobiotechnology devices. Nanobiotechnology is the development of devices built using biological molecules, cells, or components of cells. It is now possible to incorporate living cells into nanotechnology devices that carry out a variety of tasks. Researchers at Drexel University in Pennsylvania are developing **nanobiosensors**. Nanobiosensors use cells integrated into analytical machinery for the purpose of detecting specific types of molecules. These instruments can be used to run robots that float in the blood and remove microorganisms or clots. Some devices are being designed using artificial cells that can be used to deliver drugs to particular cells after being injected into the body. The artificial cells can also be used to enhance the immune system's ability to remove bacteria and viruses from the body.

REFERENCES

1. New Scientist.Com. "Ink-jet Printing Creates Tubes of Living Tissue." Available online. URL: http://www.newscientist.com/article.ns?id=dn3292.
2. Phyorg.com. "New Technique May Speed DNA Analysis." Available online. URL: http://www.physorg.com/news4201.html.
3. T. Tollefsbol, ed. *Epigenetic Protocols Series: Methods in Molecular Biology*. Totowa, N.J.: Humana Press, 2004.
4. W.C. Heiser, ed. *Gene Delivery to Mammalian Cells. Volume 2: Viral Gene Transfer Techniques*. Totowa, N.J.: Humana Press, 2003.
5. Institute of Medicine. *Informing the Future: Critical Issues in Health*. Washington, D.C.: National Academy Press, 2005.
6. E.C. Grace. *Biotechnology Unzipped: Realities*. Washington, D.C.: Joseph Henry Press, 2006.
7. B.O. Palsson and S.N. Bhatia. *Tissue Engineering*. Upper Saddle River, N.J.: Prentice Hall, 2003.
8. W.M. Saltzman. *Tissue Engineering: Engineering Principles for the Design of Replacement Organs and Tissues*. Cary, N.C.: Oxford University Press, 2004.
9. National Research Council. *Health and Medicine: Challenges for the Chemical Sciences in the 21st Century*. Washington, D.C.: National Academy Press, 2004.
10. J.P. Fisher. *Tissue Engineering: Advances in Experimental Medicine and Biology*. New York: Springer, 2006.

FURTHER READING

Books

Alberts, B., J. Lewis, M. Raff, A. Johnson, and K. Roberts. *Molecular Biology of the Cell.* London: Taylor & Francis, Inc., 2002.

Avise, J.C. *The Hope, Hype, and Reality of Genetic Engineering: Remarkable Stories from Agriculture, Industry, Medicine, and the Environment.* New York: Oxford, 2004.

Bains, W. *Biotechnology: From A to Z.* New York: Oxford University Press, 1998.

Borem, A., F.R. Santos, and D.E. Bowen. *Understanding Biotechnology.* San Francisco: Prentice Hall, 2003.

Fukuyama, F. *Our Posthuman Future: Consequences of the Biotechnology Revolution.* New York: Farrar, Straus and Giroux, 2002.

Glick, B.R., and I. Pasternak, eds. *Molecular Biotechnology: Principles and Applications of Recombinant DNA.* Washington, D.C.: ASM Press, 2002.

Jones, S. *The Language of Genes: Solving the Mysteries of Our Genetic Past, Present, and Future.* New York: Anchor/Doubleday, 1995.

Naam, R. *More Than Human: Embracing the Promise of Biological Enhancement.* New York: Broadway, 2005.

Stock, G. *Redesigning Humans: Our Inevitable Genetic Future.* Boston: Houghton Mifflin, 2003.

Thieman, W.J., and M.A. Palladamo. *Introduction to Biotechnology.* San Francisco: Benjamin Cummings, 2003.

Web Sites

Access Excellence

http://www.accessexcellence.org

Action Bioscience

http://www.actionbioscience.org

Biotechnology Institute

http://www.biotechinstitute.org

Cold Spring Harbor Laboratory

http://www.cshl.org

EurekAlert—Science News

http://www.eurekalert.org

Human Genome Project

http://www.ornl.gov/sci/techresources/Human_Genome/home.shtml

Bionanotechnology and Robotics

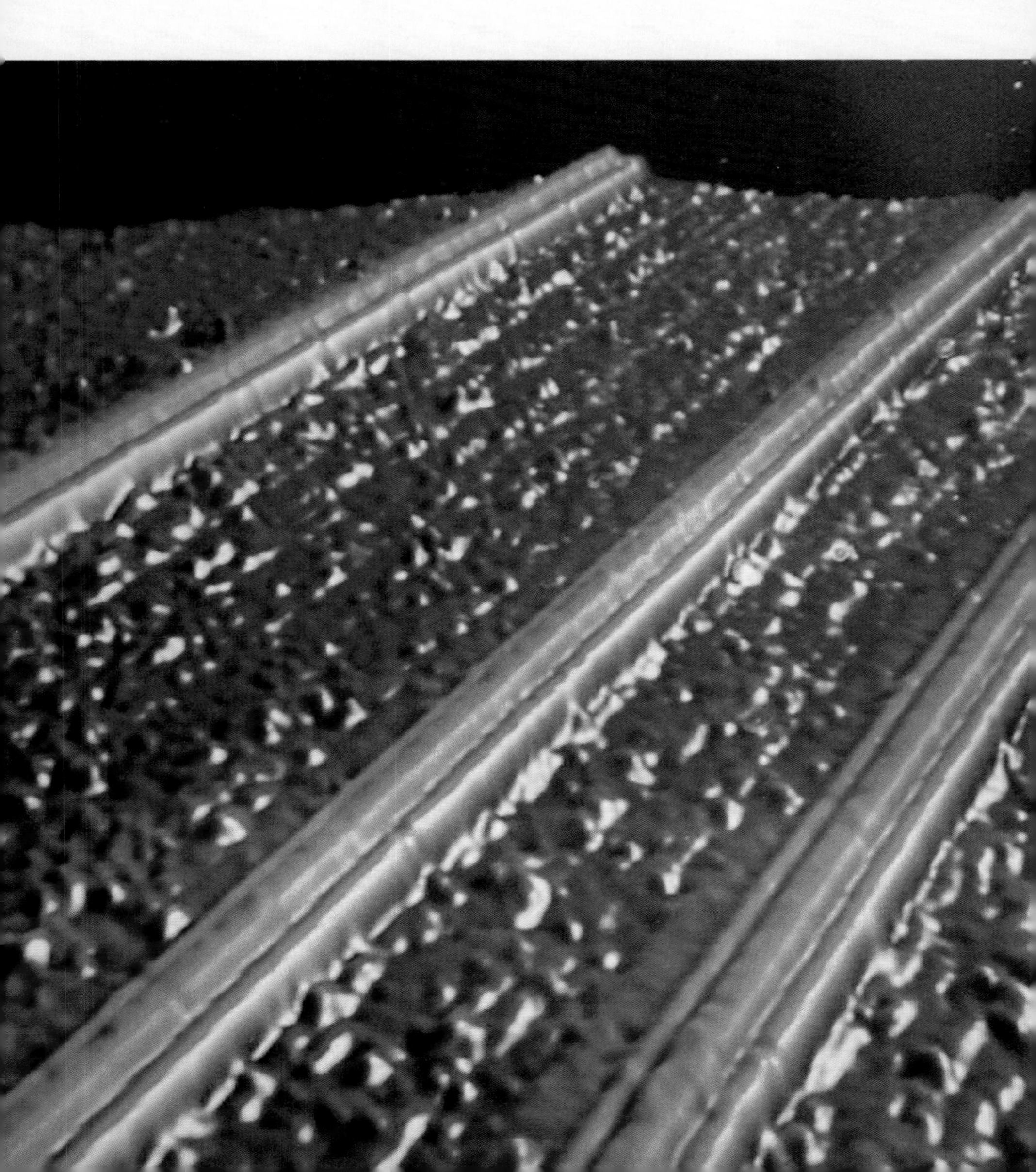

WHAT'S IN THE NEWS?

In Chapel Hill, North Carolina, another successful heart valve repair was performed on a patient in February 2005. This is not an unusual circumstance at many hospitals throughout the world. However, this surgery differed in that it was carried out by a robotic surgeon. This robot is a more versatile surgeon than most human surgeons. It is capable of performing hysterectomies, removing cancers from various body parts, repairing damaged blood vessels, and conducting gastric bypass surgery. This surgical robot called the da Vinci Surgical System (Figure 3.1) was developed and produced by Intuitive Surgical with headquarters in Sunnyvale, California, and Saint-Germain en Laye, France.[1]

The da Vinci Surgical System does not perform surgery entirely on its own. It is actually a surgical support system that greatly enhances the skills and senses of a surgeon. A surgeon has no direct contact with the patient when using the da Vinci Surgical System. He or she sees the patient by viewing the surgery through a large viewing system that provides three-dimensional imaging of the body region. The view can be greatly magnified and modified using various imaging technologies. This provides physicians with instant access to internal views and tissue details not available using traditional surgical procedures.

The patient lies on a table near the physician. Above the table is a set of large robot arms that holds the same types of surgical instruments used in traditional surgery. The surgeon controls the movements of the robotic arms and is performing the surgery on a three-dimensional image. In effect, the virtual surgery is translated into actual surgery on the patient. The arms of the da Vinci Surgical System provide accurate movements with subtle actions of the surgeon's hands. This reduces the tiring of the hands that often results from long and delicate surgeries. Robotic surgeons are catching on across the world. Surgeons in Australia are now adopting surgical robots to perform delicate surgical operations.

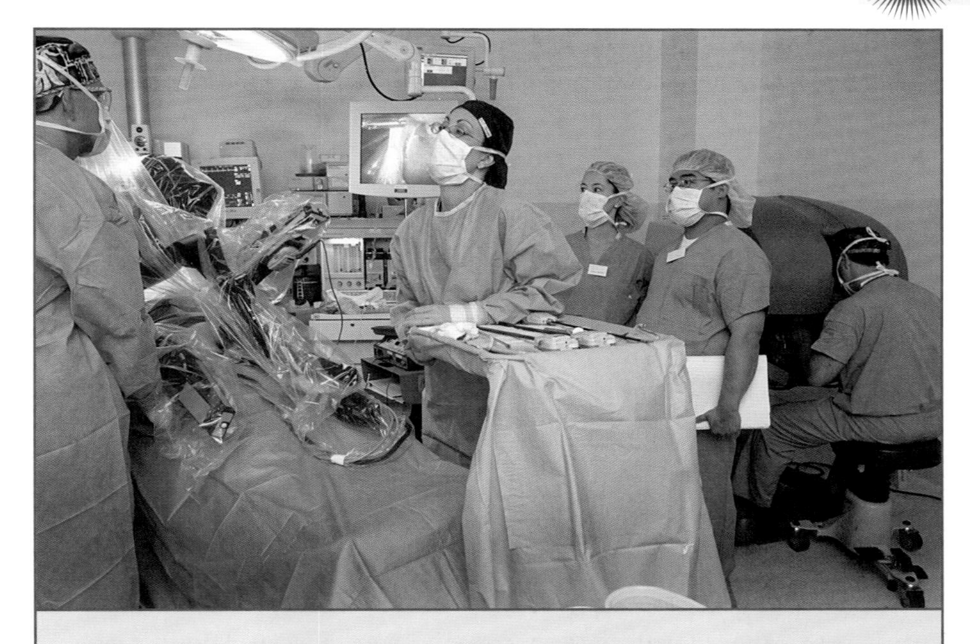

Figure 3.1 Doctors perform a gall bladder operation using the da Vinci Surgical System. The physician seated at the far right of the photo operates the system's robotic arms.

They discovered that robots permit improved precision and reduce complications following traditional surgeries. In addition, the robots reduce the amount of operating room staff needed during a traditional surgical procedure.

It is hoped that in the future surgical robots can be inserted into the body to perform meticulous microsurgical procedures without opening up the body. Scientists at the Swiss Federal Institute of Technology in Zurich, Switzerland, have created a prototype robot for this task. It is the smallest robot of its kind and is made of microscopic components. The "biomedical microrobotic system" can be propelled through maze-filled water designed to model a trip through a blood vessel or a body cavity. Small magnets help orient the robot. The researchers hope to use the robot for noninvasive repair of blood vessel and heart damage. Similar robots can be used to swim around the blood removing cancerous cells and invasive organisms. It is also possible to use these robots to monitor vital body functions in a particular tissue or organ.[2]

INTRODUCTION

Many technological innovations started out as large, cumbersome devices only to evolve into miniature tools that look more like toys than complex mechanisms. A prime example is the earliest of the electronic high-speed computers. One computer, called the Harvard Mark I computer, was developed in 1944 by Howard Aiken and Grace Hopper. It weighed 10,000 pounds (4,536 kilograms) and measured 55 feet (16.7 meters) long and 8 feet (2.4 m) high. The computer comprised approximately 760,000 separate pieces. A smaller computer called the Electrical Numerical Integrator and Calculator (ENIAC 1) was developed for the military in 1946 by John Mauchly and J. Presper Eckert. This computer took up 1,800 square feet (167.2 square meters) of floor space and used approximately 10,000 capacitors, 1,500 relay switches, 70,000 resistors, and 17,000 vacuum tubes.

Advances in an understanding of the chemistry and physics of electric circuitry led to the miniaturization of computers. The same computing power of Mark I and ENIAC 1 was packed into the IBM PC, introduced in 1981, and the Apple desktop computer, released in 1977. Today, it is possible to get hundreds of times the capabilities of these earlier "small" computers into a chip the size of a pinky fingernail. Computers are not the only shrinking technologies. Mechanical devices used in a variety of purposes ranging from clocks to electrical switches are being reduced to microscopic sizes. Robots are the most popular mechanical device undergoing miniaturization. The term *robot* usually refers to any mechanical device that carries out some of the functions of a human or an animal and that moves automatically or by remote control. It also includes any programmable device that manipulates materials, parts, or tools using controlled repetitive motions. Robots are used for a variety of purposes, including tasks such as assembling automobiles, packaging foods, dispensing samples of blood to a medical analysis instrument, and orienting the camera on a space exploration satellite.

The first device defined as a robot was developed around 270 B.C. by the Greek engineer Ctesibus. He invented elaborate musical organs and water clocks decorated with animated figures. The term *robot* was created by Czech writer Karel Capek for a play called *Rossum's Universal Robots* written in 1921. Scientist and writer Isaac Asimov coined the term *robotics* in 1941 for a series of science fiction stories. In 1948, the science **cybernetics** was born. Cybernetics is the use of artificial intelligence to run robots. By the late 1950s, simple robots were being used in industry for assembling and manufacturing various commercial products. Most of these earlier robots were simple robotic arms that carried out a limited selection of movements. Improvements in computers and electrical circuitry during the 1980s and 1990s led to the development of robots capable of carrying out many tasks. Many of these robots were used for exotic purposes such as bomb disposal and space exploration. By 2000, robots were making their way into households. Today, there are robotic kitchen appliances and vacuum cleaners in many houses.

The science of nanotechnology is exploiting the lowest possible limits at which computers and machines can be designed. The *nano-* root term in *nanotechnology* means dimensions in the size range of billionths of a linear or volume measurement unit. **Bionanotechnology** is a specialized component of nanotechnology that uses biological materials or biological principles to design technologies for biological or medical applications. This new science is giving robots new types of jobs that cannot be done by the earlier robots. It is conceivable that bionanotechnology innovations will become common household items just like computers and robot appliances.

WHAT IS BIONANOTECHNOLOGY?

Bionanotechnology is the creation of nanotechnology devices using biological molecules. Nanotechnology is acclaimed by the National Aeronautics and Space Administration as one of the most exciting emerging technologies of this century. Nanocrystals, nanofibers, nanotubes, and nanowires are being discussed as inventions that would greatly advance medical diagnosis and treatments. Microscopic tubular structures are being developed as the preferred way to carry minute currents and voltages in electronic devices implanted in the body or placed in small handheld medical instruments. Nanotechnology seemed to have just appeared in the 1990s with claims in popular magazines about a new technology for building incredible microscopic robots. Few people are aware that the fundamental ideas of nanotechnology date from the birth of contemporary genetics.[3]

In 1959, physicist Richard Feynman made what is now considered an epic statement at a meeting of the American Physical Society held at the California Institute of Technology. He stated, "What I want to talk about is the problem of manipulating and controlling things on a small scale." This was the first serious discussion about the chemistry and physics of designing infinitesimally small electronic and mechanical devices. Feynman amazed the audience of accomplished physicists with his critique of the technological potential and limitations of designing microscopic devices. This talk spurred numerous investigations into the miniaturization of various electronic components and mechanical equipment. By 1974, physicist Norio Taniguchi of Tokyo Science University first used the word *nanotechnology* to describe the scale of size discussed by Feynman.

Nanotechnology moved from a theoretical concept to an actual technology because of the research of MIT engineer Kim Eric Drexler in 1977. He spurred the growth of molecular nanotechnology that focused on the chemistry of materials that can be used to design microscopic devices. He dismantled many of the barriers

Figure 3.2 The discovery of the buckyball paved the way for the development of bionanotechnology devices. In the photo above, Harry Kroto holds a model of the buckyball. Kroto, along with Richard Smalley, received the 1996 Nobel Prize in Chemistry for his discovery of this fascinating molecule.

that inhibited the practicality of nanotechnology. Drexler defined the engineering needs for nanotechnology by comparing the components of traditional large-scale technologies to the necessary parts needed for equivalent microscopic technologies as shown in the table on page 64. Note that many of the nanotechnologies are composed of biological materials such as **microtubules** from cells and a variety of molecules comprising cells and tissues. Drexler's paper headed nanotechnology into the bionanotechnology direction.

Drexler's ideas were turned into reality in 1985 with the creation of a structure called buckminsterfullerene or the "buckyball." This miniscule object was a cage formed of 60 carbon atoms molded into a hollow ball (Figure 3.2). It is named in honor of Buckminster

Fuller, who created architectural structures called geodesic domes. Harry Kroto and Richard Smalley of Rice University discovered the first buckyball. Their research led to the development of microscopic levels and tubes made of biological molecules that now form the foundation of modern bionanotechnology. The research team received the 1996 Noble Prize in Chemistry for their work on buckyballs and other bionanotechnology investigations.

In 1989, a device called the scanning tunneling microscope (STM) was developed as a new tool for nanotechnology work. It is a microscope that uses a tiny needle that nearly touches the atoms making up the surface of objects too small to be detected with traditional microscopes. Researchers working independently at IBM and the Massachusetts Institute of Technology used the STM to move atoms around for the creation of nanotechnology components. They respectively moved atoms around on the surface of an object to spell out IBM and MIT. This use of the invention seemed frivolous; however, it supported the hypothesis that scientists could use STM as a tool for building bionanotechnology devices out of a variety of biological materials.

Discussions from the first nanotechnology brainstorming meeting called "The Potential of Nanotechnology for Molecular Manufacturing" were published in a monograph by the Rand Research Group in 1995.[4] This document set the direction for future bionanotechnology research. The report recognized that the existing nanotechnology devices were promising technologies for further development. Many of the barriers that repressed the development of bionanotechnology devices were being torn down. The promising forecast outlined in the report compelled NASA and U.S. military sciences to pursue bionanotechnology investigation. It also promoted the creation of bionanotechnology research laboratories at major universities throughout the world. By 1998, a wide array of prototype bionanotechnology devices such as simple microscopic robots and DNA-based neural networks were being developed.

In 2005, the international Nano Ethics Conference, held at the University of South Carolina, discussed the implications of more than 300 nanotechnology devices currently under development.

At the time of this conference, bionanotechnology included the molecular manufacturing of devices ranging in uses from microscopic medical instruments that provide real-time health monitoring to miniature self-replicating robots capable of selectively damaging an enemy nation's natural resources. Specialty fields of bionanotechnology are forming as new applications become evident with advancing technologies. Current research fields such as artificial intelligence (AI) are being redirected into bionanotechnology direction. Bionanotechnology is considered the future of the computer and electronics industries.[5]

CURRENT AND FUTURE USES OF BIONANOTECHNOLOGY

Bionanotechnology research is making rapid advances and will likely be replacing many traditional technologies. A major limitation of bionanotechnology is that the current instruments used to manipulate molecules do not have the sensitivity required to accurately or effectively determine the characteristics of the resulting devices. However, this is not impeding the design of theoretical machines that are studied to the best of the current capabilities. This gap in technology is spurring scientists to develop new types of microscopes and analytical instruments that detect the structure and function of bionanotechnology devices.

DNA and proteins are the most common building materials selected for bionanotechnology devices, or **bionanomachines,** today. Proteins can be molded into a variety of flexible and rigid structures that are durable under a variety of environmental conditions. Most of the proteins being developed for bionanomachines do not exist in nature. They are synthesized by binding **amino acids** into chains that take on a variety of properties based on the amino arrangement and composition. The current technologies driving contemporary and future bionanotechnology applications include creative ways of adapting cell functions into the operation of microscopic machines.[6]

Traditional Technology versus Nanotechnology

Traditional Technology	Nanotechnology Equivalent	Function
Struts, beams, casings	Microtubules, cellulose, mineral structures	Transmit force, hold positions
Cables	Collagen	Transmit tension
Fasteners, glue	Intermolecular forces	Connect parts
Solenoids, actuators	Conformation-changing proteins, actin/myosin	Move things
Motors	Flagellar motor	Turn shafts
Drive shafts	Bacterial flagella	Transmit torque
Bearings	Sigma bonds	Support moving parts
Containers	Vesicles	Hold fluids
Pipes	Various tubular structures	Carry fluids
Pumps	Flagella, membrane proteins	Move fluids
Conveyor belts	RNA moved by fixed ribosome	Move components
Clamps	Enzymatic binding sites	Hold workpieces
Tools	Metallic complexes, functional groups	Modify workpieces
Production lines	Enzyme systems, ribosomes	Construct devices
Numerical control systems	Genetic system	Store and read programs

Adapted from: K. Eric Drexler, "Molecular Engineering: An Approach to the Development of General Capabilities for Molecular Manipulation." *Proc. Natl. Acad. Sci.* Vol. 78, No. 9 (1981): 5275–5278.

One typical type of modern bionanomachine will have a body molded of protein components held together by the same types of **hydrogen bonds** that help link together many body components. Hydrogen bonds are a strong force at microscopic levels and will serve as glue for attaching different pieces of the bionanomachines. Simpler bionanomachines are formed by attaching carbon atoms into buckyball or microtubule shapes. Microtubules are hollow microscopic tubes. Buckyballs filled with drugs have already been outfitted with other molecules that help them stick to cells for use in targeted-drug delivery systems. They act as microscopic **syringes** that can inject the contents of a buckyball into a specific cell to which they selectively attach.

Moving parts will be driven by enzymes such as **dynein**, which is a motor protein that can move across the surface of other materials. These dynein motors will do a variety of tasks, including moving propeller-like objects and small levers in microscopic robots. The fuel for the dynein is a molecule called **adenosine triphosphate**, or ATP. ATP is a molecule that transfers energy from one molecule to another. Helen C. Taylor and Michael E.J. Holwill of the physics department at King's College London have investigated many bionanotechnology applications of whole cell units called axonemal motors, which drive cell structures called **cilia** and **flagella**. Cilia are short, movable, hairlike structures found on the surfaces of some types of cells. Flagella are long, hairlike projections used for movement in some microorganisms and cells. Taylor and Holwill found it possible to build axonemal motors into microscopic robots.

Scientists in the military already envision that axonemal robots will carry out tasks such as finding and disabling land mines. These microscopic robots could move through the soil and find objects containing explosives. They could then be outfitted with radio **global positioning system** (GPS) devices that signal the location of the ammunition or they can carry bacteria known to break down the chemical components of the explosive materials. GPS is a satellite-based radio positioning system that identifies the location of an object on Earth. Prototype axonemal robots are being developed that could swim through the blood to deliver drugs or surgically remove a blood clot in a blood vessel or the heart. It is also being proposed to use axonemal robots that supplement the immune system by attacking **parasites** that cannot be adequately treated with conventional medicine.[7]

Another current application of bionanotechnology is the DNA computer. These microscopic and complex devices were first developed in 1994 by Leonard Adelman of the University of Southern California. He spliced DNA and recombined the pieces into patterns that resembled flowcharts for solving mathematical formulas. Adelman then used a biotechnology technique called **electrophoresis** to find the simplest pattern or pathway. Electrophoresis is the use of electricity to separate different types of nucleic acids and proteins. This device was designed to model a **Turing machine** (named after British mathematician Alan Mathison Turing). Turing machines, which were first developed in 1936, are simple computational devices designed to investigate the simplest path to determine a mathematical calculation. In effect, Adelman's DNA computer was similar to searching for the shortest route of highways and roads between two distant points. It quickly gave "true" or "false" answers that pointed to the simplest way to determine the answer to a problem. Researchers hope to use DNA computers that are microscopic programs for machines that identify ailments when injected into the body.

Another application of bionanotechnology is artificial life forms. They are based on the discovery of **protocells** that were described and created by Sidney Fox at Southern Illinois University at Carbondale and the University of Miami in the 1970s. These simple cells were made of bubbles of fats that surrounded a mixture of chemicals resembling certain metabolic pathways of living organisms. Research laboratories throughout the world were able to get protocells to carry out a variety of cells that could replace the missing functions in diseased cells. One protocell was able to produce energy and specific chemicals using photosynthesis much like a plant.

Protocells currently have not been used for any commercial or medical applications. However, they are useful in understanding the development and evolution of different types of cells. This information may lead to the ability to produce artificial muscle and nerves that can form the base of transplants and robotic devices.[8] Simple protocells called liposomes were used by researchers at University of California at Los Angeles to transfer chemicals and genes into diseased cells.[9] Liposomes are defined as microscopic spheres of fat that enclose DNA or medication. In 2003, a research team designed a liposome that fused to brain cells and transferred genes into the nerve cells of the brain. It is hoped that this method can be used to treat genetic disorders of the brain, including Parkinson's disease. This application of artificial cells is called gene therapy.

A technology related to artificial cell gene therapy involves the production of self-replicating machines that resemble viruses. Scientists are able to build a protein capsule that adheres to specific cells in the body. This capsule resembles the primary structure making up a virus. Inside the artificial virus are correct copies of genes or medications used to treat disease. On the surface of the capsule are proteins that entice the cell to take in the artificial virus and its contents. Scientists hope to build self-replicating artificial viruses that carry medicines or vaccines in the body, which would last for the duration of a person's life.

WHAT IS ROBOTICS?

Mentioning the term *robot* to many people conjures thoughts of the types of mechanical devices seen in movies and TV shows. A robot is defined as any mechanical device or computer program that resembles a living organism and operates automatically or by remote control. The typical robots of *Bicentennial Man, Lost in Space,* and *Star Wars* are fanciful ideas that may one day find a place in society. Robotic devices have been making their way into everyday life since the 1960s with the first talking and walking dolls that became popular toys throughout the world. This field of robotics is called consumer robotics. Robotic toys became increasingly sophisticated, leading to virtual pets that "die" if not cared for properly. Animatronics is a technology in which robotic toy devices are designed to create large, animated, virtual organisms for the entertainment industry. Many of these creations can be seen in popular children's restaurants and at theme parks.

Consumer robotics has not been limited to toys and entertainment devices. Many robots have been designed to assist with various chores around the house and office. In September 2003 a robotic vacuum cleaner called the Roomba® was developed by iRobot in Boston, Massachusetts. A small, disk-shaped machine moves around the floor vacuuming the house. It uses a simple sensor and computer that permits it to clean the floor. A similar device, RoboMower®, was developed by Friendly Robotics to mow yards. These simple robots use scaled-down technologies of the ones used to run more complex robots such as the NASA Mars Rover.

A report produced in 2004 by the United Nations Economic Commission for Europe (UNECE) and the International Federation of Robotics (IFR) showed that robots have a promising future in various commercial and medical uses.[10] The worldwide investment robot technology has been increasing approximately 20 percent per year between 2002 and 2004. According to the report, this growth is predicted to continue through 2007. The report showed that more than 600,000 household robots were being used globally in 2003. Robots are not a passing fad, and they are not solely destined to be a novelty for carrying out simple chores. Robots with many medical applications are already in use. Many new robots are being designed to assist medical procedures.

Robotic devices are based on one or more machine design principles. **Articulated** robots are composed of levers and pulleys that carry out movements resembling human appendages such as hands or legs. These robots are commonly used for assembling commercial items using humanlike motions. Articulated robots can be developed to hold large welding torches or delicate surgical instruments. They use rotational joints to maneuver the levers. Cartesian or gantry robots slide structures along a series of tracks that guide the movement of the machine. The term **Cartesian** is named after the seventeenth-century philosopher René Descartes and refers to the mathematics of movements and positions. They are used for positioning items or other robotic parts over a specific location. Computer and location sensors are used to move the gantry device into a precise position. These robots can be seen placing the caps on glass or plastic bottles in soda-bottling companies. Articulated and Cartesian robots can be designed into complex robotic machines.[11]

Cylindrical robots are composed of an arm that swings in a circular direction as seen in the rotational movement of R2D2 in the *Star Wars* movies. These robots are used to move objects from one location to another. For example, these robots are found in sorting machines that pick up and dispose of defective items on a conveyer belt. Spherical or polar robots can rotate around a pivot in a three-dimensional motion, allowing them to move objects up and down or in a circular motion to twist tools such as screwdrivers, spot welders, and wrenches. Computers control the movement, and they can be attached to articulated and Cartesian robots. They combine the properties of cylindrical and Cartesian robots.

SCARA or Selective Compliant Articulated/Assembly Robot Arm robots are composed of two or more arms controlled by pivot or rotational joints. They are used for repetitive jobs involving picking up and placing parts. These robots provide movement in what is called a compliance plane. This means that they move with the precision of a human arm and combine the properties of articulated and spherical robots. Parallel robots use movable pistons to move a mobile platform. These robots are used in flight simulators to provide all the rocking and twisting movements of a real airplane. They are also used in amusement park rides that create the feel of driving or flying. Parallel robots can be fashioned with computers that help level industrial and medical machinery that must be keep steady.

The movements of robots are manipulated by devices called robotics control systems (RBS). An RBS can be controlled by human movements or by computer software designed for the system. The heart of the RBS is the intelligent navigation coordination control system. This device synchronizes the robot's movements to the commands of the human controller or the computer software. The intelligent navigation coordination control system communicates each movable component of the robot through standard IEEE (Institute of Electrical and Electronics Engineers) attachments or a wireless Ethernet communication system. Standard IEEE attachments plug directly into the computer. A wireless Ethernet communication system uses radio waves to operate the robot remotely without a direct connection. The IEEE or Ethernet systems send an electrical signal through a network of circuits and wires that operate motors in the robot. The motors control the actions of the arms or other movable devices.[12]

The movable parts of the robot are outfitted with **servos** that provide movement and sensors that detect various aspects of the robot's movements. A servo is a small, motor-driven device that has an output shaft. This shaft can be attached to a robotic device to produce specific angular positions. Sensors send electrical signals to the RBS. This process provides important feedback needed to direct the movements of the robot. Sensors can be designed to tell the position or angle of movement. There are also sensors that tell the robot the force at which it is working. These two types of sensors can be combined to provide the same range of movement and sensitivity to touch found in a human hand. Sensors can also be designed to tell the robot information such as chemical composition, color, temperature, and texture. Laser sensors are commonly used to guide the movement of robots. Many robots contain various types of cameras that send images to the human controller or to the computer.

CURRENT AND FUTURE USES OF ROBOTICS

Biophysicist and science fiction author Isaac Asimov predicted one future direction of medical robotics in his 1976 book *The Bicentennial Man.* Asimov realistically illustrated the potential for science to produce robotic parts that modeled the function of body organs. Unlike those in the book, today's medical scientists are not using robots to build an artificial human, or cyborg. Rather, the technology is replacing ailing body components in a

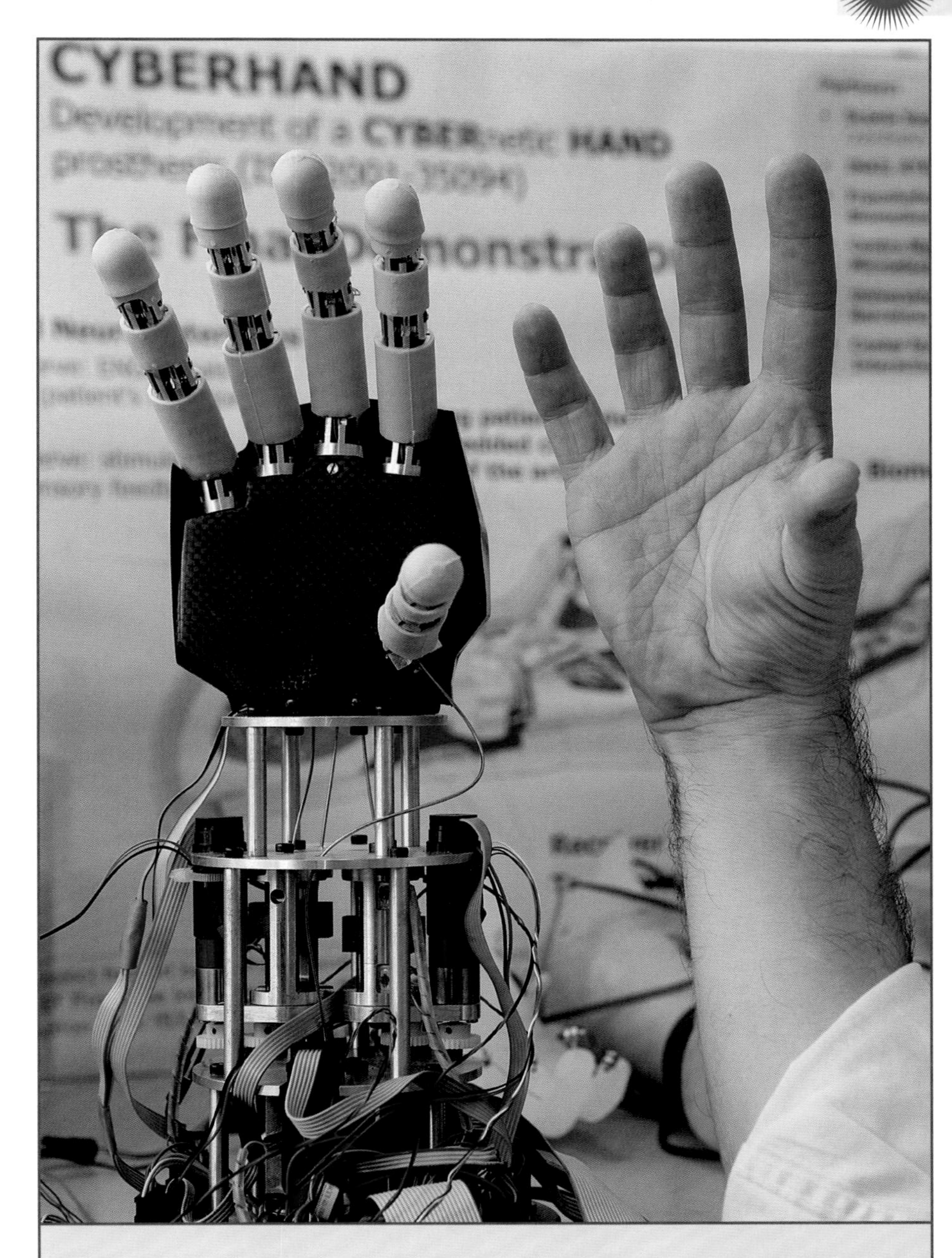

Figure 3.3 A model of the Cyberhand is photographed next to a human hand at an engineering laboratory in Italy. Currently in the early stages of development, Cyberhand is a prosthetic hand that will be able to re-create the sensation of touch.

high-tech application of prosthesis. The traditional practice of prosthesis has changed greatly with the invention of new materials and robotic devices for operating artificial limbs and organs. Earlier prosthetic devices were simple machines with simple programming that replaced what are known as motor functions of the body. Motor functions are the actions of neural signals sent from the brain to a muscle or organ. Newer devices are capable of replacing sensory body functions (Figure 3.3). Sensory functions are neural signals that communicate information about the environment to the brain. Balance, hearing, pain, sight, smell, and touch are forms of sensory information.

The common types of medical robotics used in prosthesis today are artificial limbs, cardiac prosthetics, cochlear implants, neuroprosthetics, ocular prosthetics, replacement joints, and somato prosthetics. They are designed according to a new scientific principle called **biomimetics**. Biomimetics, also called bionics or biognosis, is the design of computers and machines by mimicking the natural movements and thought processes of living organisms. It involves the cooperation of psychologists as well as engineers and scientists. Frank Grasso, of Brooklyn College in New York City, is a psychologist who studies the programmed behaviors of lobsters for use in programming robotic computer programs. These programs can then be used to coordinate the movement of artificial limbs. This makes the artificial limbs move with the same natural motions and responses of actual limbs.[13]

Artificial limbs and artificial joints are used to replace severely damaged or lost components of arms and legs. The prosthetic robots used in traditional prosthetic limbs are powered by a battery and controlled by **analog** circuits that respond to the twitching of muscles fitted to the artificial arm or leg. Analog circuits involve the transmission of a continuous wave electrical signal that varies in amplitude in response to specific types of input such as muscle twitches. These devices can be heavier than an actual limb because of the use of metallic supporting and moving parts. It takes much training through numerous physical therapy sessions for the patient to use the analog controls of the limb effectively. New versions of the traditional robotic limbs are made of lighter-weight materials and require less energy to operate. Plus, the new joint materials do not wear out as easily as metal and provide smoother, more natural movements. Robotic limbs being developed now are using a variety of new technologies that make them as usable as an actual limb.

A prosthetic limb called the Boston Digital Arm System (BDAS), developed by Liberating Technology Incorporated in Massachusetts, uses digital technology that provides better movement and more sensitive controls. A BDAS hand can easily grip an egg without cracking it while also permitting the user to make a strong grip that can support the weight of the body. It is operated by sensors called myoelectrodes that detect subtle differences in muscle action. Unlike analog controls, these sensors control the limb the same way a real limb responds to rapid electrical variations associated with light and strong muscle contractions. It also uses what the company calls "Touch Pads" that act like a nervous system that tells the control how much tension or pressure is being transmitted to a movement. The limbs use less electricity than the older technologies and have built-in computers with software for carrying out and learning routine tasks. BDAS limbs have been tested on various household and workplace activities including all the tasks of driving a car.

Cardiac prosthesis ranges from heart-assist devices to heart replacements. Heart-assist devices are more of an electrical signaling device than a robot. These devices, called pacemakers, serve as an artificial brain that sends electrical signals that keep the heart beating consistently. Older devices did not adequately adjust to changes in activity and breathing. Newer pacemakers have microcomputer chips that can monitor various body conditions and accordingly adjust the heartbeat. Robotics is making a greater impact on artificial hearts. Traditional artificial hearts were machines that only replaced the valves of the heart. An external pump and controlling device operated the heart and kept the heart in pace with the rest of the body. New hearts such as the CardioWest temporary Total Artificial Heart (TAH-t) by SynCardia Systems, Inc., still require external control units. However, it pumps internally like a real heart and it gives the patient more mobility because it uses computer software to control the robotic valves. The first internal pumping heart was the Jarvik 7 developed by Robert K. Jarvik in the early 1980s. It is hoped that newer robotic hearts will have internal energy on control units that give a patient the appearance of having an actual heart. New biomaterials may make it possible for the artificial replacement hearts to have the same durability and longevity of an actual heart.

Computerized robots for replacing body senses include cochlear implants, neuroprosthetics, and ocular prosthetics. These devices use transducers and microcomputers to convert light, pressure, sound, or touch into appropriate neural signals that replicate body senses. Most contemporary artificial hearing devices either amplify sound or conduct sound to the brain indirectly by tapping on bones near the ear. These devices do not provide a full range of hearing and do not work in people who have nerve damage. In 2005, scientists at the University of Michigan built the first life-size artificial cochlear. The cochlear is the transducer of the ear that converts sound into nerve signals. The researchers hope that it can be wired into the nervous system in people with hearing loss due to nervous system damage.

The first attempt at an artificial eye was a specialized miniature camera tested in 2005. Researchers at the Doheny Retina Institute of the University of Southern California wired a special camera to the visual area of the brain. Signals from the camera were converted into crude images that replaced basic image perception. This camera may ultimately be outfitted into the first robotic eye, developed in 2000. Further knowledge of brain function may permit scientists to build artificial neural pathways that fully integrate robotic ears and eyes into the brain's circuitry. An artificial larynx or voicebox is being investigated by researchers at the University of Waterloo in Canada.

Other robotic body parts are being developed to replace major body organs such as the kidney, liver, and pancreas. These somatoprostheses are currently external devices that limit patient mobility. Implantable pumps are being used to replace the function of glands that secrete chemicals into the blood. Many of these devices still require an external power source or a means of injecting the

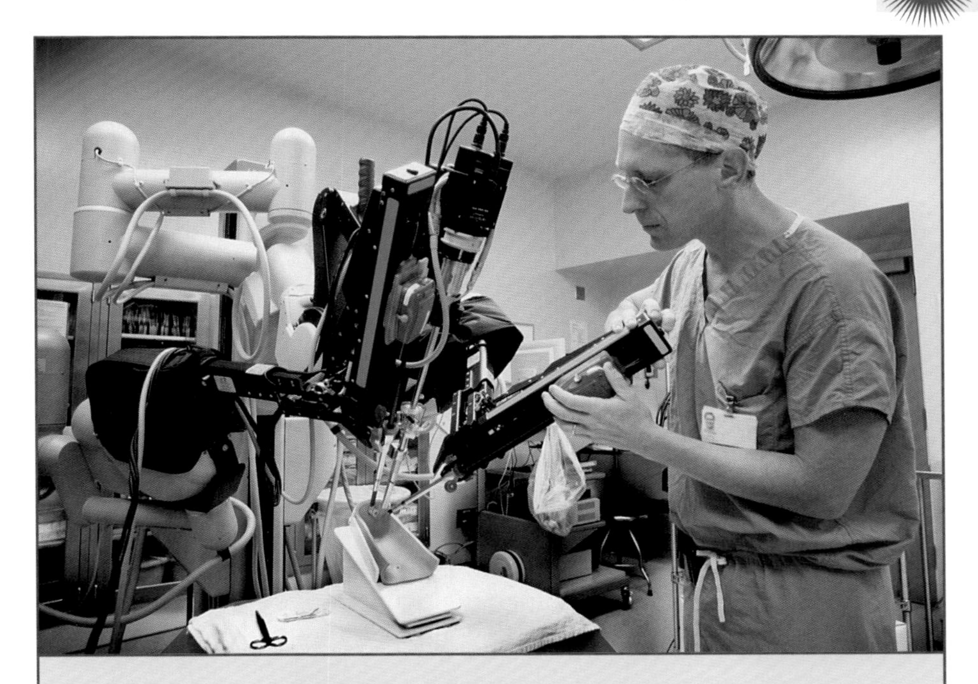

Figure 3.4 A doctor adjusts the arms of a surgical robot. The state-of-the-art robot is used for heart, prostate, and other surgical procedures.

chemicals into the device. Scientists at the Department of Energy's Lawrence Livermore National Laboratory and MiniMed, Incorporated, in California are developing a new style of artificial pancreas to help diabetics. This device uses lasers to monitor the body's conditions and deliver the correct amount of chemicals to keep the body in balance. It is hoped that these devices can be combined with living cells that work together with the robotic components.

Another area of medical robotics assists physicians with the diagnosis and treatment of injury and diseases. Medical robots called master-slave robotic systems are becoming a popular way to improve medical procedures. These systems are controlled by a physician (the master) who manipulates a panel or joystick to control a robot (the slave), which carries out a medical procedure (Figure 3.4). Researchers at Eindhoven University of Technology in the Netherlands recently developed a prototype medical robot that helps a surgeon perform delicate surgeries and diagnostic procedures. The robot converts large hand motions into delicate patterns of movement needed for working on nearly microscopic surgeries. The robot permits the surgeon to do complex procedures through a very small incision. Miniature cameras and sensors act as eyes that can see into the tissues. The scalpel is replaced by a device called a **plasma** needle. Plasma is an extremely hot gas that is composed of charged particles. A plasma needle can cut and seal microscopic sections of tissue with little need to control bleeding or suture the cut tissues.[14]

Similar robots are being outfitted with diagnostic instruments that can carry out many medical measurements without the need for large, invasive incisions. Laser attachments to robots are being developed that can test blood chemistry without the need for collecting blood. Other detectors, such as **Doppler**, **ultrasound**, and other types of imaging, are being developed to monitor blood pressure and fat buildup in blood vessels. Electrical sensors attached to robots may be able to replace traditional **electrocardiograms** and **electroencephalograms**. Computer software in the robots will even help diagnose the condition. Remote control medical robots are already being investigated at the French National Institute for Computer Science and Control in which physicians can guide medical robots from distant locations. It is even hoped that, eventually, each household will have simple-to-use medical robots for diagnosing common ailments and for performing minor medical procedures.

REFERENCES

1. University of California Keck School of Medicine. "Robotic Surgery Institute." Available online. URL: http://www.cts.usc.edu/rsi-davincisystem.html.
2. Institute of Robotics and Intelligent Systems. Available online. URL: http://www.iris.ethz.ch/msrl/.
3. E. Roduner. *Nanoscopic Materials: Size Dependent Phenomena.* London: Royal Society of Chemistry, 2006.
4. M. Nelson and C. Shipbaugh. *The Potential of Nanotechnology for Molecular Manufacturing.* Santa Monica, Calif.: Rand Foundation, 1995.
5. C.J.M. Van Rijn. *Nano and Micro Engineered Membrane Technology.* Oxford: Elsevier Science, 2004.
6. C.M. Niemeyer and C.A. Mirkin, eds. *Nanobiotechnology: Concepts, Applications and Perspectives.* New York: John Wiley and Sons, 2004.
7. E. Straus. *Medical Marvels.* Amherst, N.Y.: Prometheus Books, 2006.
8. J. Chela-Flores and Francois Raulin, eds. *Chemical Evolution: Physics of the Origin and Evolution of Life: Proceedings of the Fourth Trieste Conference on Chemical Evolution, Trieste, Italy, 4-8 September 1995.* New York: Springer, 1996.
9. J. Huwyler, D. Wu, and W.M. Pardridge. "Brain Drug Delivery of Small Molecules Using Immunoliposomes." Proceedings of the National Academy of Sciences. Vol. 93 (1996): 14164–14169.
10. United Nations Economic Commission for Europe. *World Robotics 2004.* Geneva, Switzerland: United Nations, 2004.
11. L. Sciavicco and B. Siciliano. *Modeling and Control of Robot Manipulators.* New York: Springer, 2000.
12. J.O. Gray and D.G. Caldwell. *Advanced Robotics and Intelligent Machines.* London: Institute of Electrical Engineers, 1996.
13. M.H. Ang, Jr., and O. Khatib. *Experimental Robotics IX: The 9th International Symposium on Experimental Robotics.* New York: Springer, 2006.
14. J. Troccaz, E. Grimson, and R. Mösges, eds. *CVRMED II and MRCAS III, Lecture Notes in Computer Science.* New York: Springer, 1997.

FURTHER READING

Books

Cobb, A.B. *The Bionic Human.* New York: Rosen Publishing, 2003.

Edwards, S.A. *The Nanotech Pioneers: Where Are They Taking Us?* New York: John Wiley and Sons, 2006.

Furmento, M. *BioEvolution: How Biotechnology Is Changing the World.* San Francisco: Encounter Books, 2003.

Goodsell, D.S. *Bionanotechnology.* Hoboken, N.J.: Wiley-Liss. 2004.

Hall, J.S. *Nanofuture: What's Next for Nanotechnology.* Amherst, N.Y.: Prometheus Books, 2005.

Ratner, D. and M.A. Ratner. *Nanotechnology: A Gentle Introduction to the Next Big Idea.* Upper Saddle River, N.J.: Pearson Education, 2002.

Williams, T.I. *A Short History of Twentieth Century Technology.* New York: Clarendon Press, 1992.

Web Sites

BioCom
http://www.bio.com

BioTech: Life Sciences Resources and Reference Tools
http://biotech.icmb.utexas.edu

Nanotechnology at Scientific American
http://www.sciam.com/nanotech

National Center for Biotechnology Information
http://www.ncbi.nlm.nih.gov

Section 4

Pharmacogenetics and Precision Vaccines

WHAT'S IN THE NEWS?

Drug development is a time-intensive and expensive process, usually only affordable to large pharmaceutical companies. Much of the money to develop a drug goes into the research and development (R&D) and clinical testing. R&D involves investigations using computers that perform chemical modeling and laboratory studies on animals and cell cultures. Years of mandatory clinical testing on both animals and humans ensure that the prospective drug is effective and will not cause dangerous side effects. Recent advances in biotechnology are now making it possible for drugs to be developed at less cost. In 2006, the first national conference on **immunoinformatics**, held at the Sri Ramachandra Medical College in Porur, India, provided hope for developing nations wishing to join the drug-development market.[1] Conference participants presented many ideas that could improve health care at a low cost in developing nations.

Immunoinformatics, which is a subdivision of a larger field known as **pharmacogenetics**, uses computers to study how the immune system functions in response to disease and drugs. It can also determine how genetics affects the outcome of a disease and its subsequent treatment. Immunoinformatics researchers commonly investigate the chemical effects of an **epitope** on the immune system. An epitope is the part of a chemical that stimulates an immune response. Epitopes are usually common components of complex carbohydrates and proteins. Any chemical that is recognized by the immune system is called an antigen. Epitopes are the part of the antigen that determines the way the body responds to the antigen. Sometimes the response leads to conditions such as allergies or rheumatoid arthritis.[2]

Other immunoinformatics conferences such as the 2005 International Immunoinformatics Symposium in Boston, Massachusetts, saluted the potential impacts of this emerging science. The bioinformatics tools needed to pursue immunoinformatics research, however, are still in their infancy. There is only a limited knowledge of the data interpretation needed to design computer software to work with a wide array of diseases. In addition, the complexities of the immune system's responses to particular epitopes are still being researched.

INTRODUCTION

Targeted medicine is the new buzzword among medical practitioners and scientists such as pharmacologists who help develop new therapeutic treatments for disease. The term *targeting* means developing treatments that specifically treat the cause of an ailment without having negative effects on the body. Historically, medical treatments have treated the symptoms of a disease, without curing it. This approach is still taken today. For example, many remedies for the common cold are a broad group of drugs called nonsteroidal anti-inflammatory medications. They simply reduce the production of a chemical called histamine that causes tissues to swell with blood and increases mucus production. In effect, the cold is not cured. However, the person feels better because some of the symptoms that make him or her feel ill are temporarily alleviated.

Modern targeted medicine relies on the understanding of how genetics affects the cause, development, and management of a disease. It relies on the investigations into **bioinformatics**, **environmomics**, **genomics**, **metabolomics**, and **physiomics**. Bioinformatics is the collection, organization, and analysis of large amounts of biological data, using networks of computers and databases. Genomics involves the study of an organism's complete DNA information. Environomics is the science investigating the role of the environment and drugs on the expression of genetic material. Metabolomics is the investigation of how genes affect the metabolism of a cell. Physiomics is the study of the genetics of metabolic functions in the body. All these studies must be carried out at the same time as research that investigates the chemistry of drug interactions with cells.[3]

The first targeted medicines fit into two major therapeutic categories: chemotherapy and **vaccination**. Chemotherapy is the treatment of disease using poisonous chemicals that kill cells. Vaccination is a therapeutic process by which a person's immune system is induced to develop protection from a particular disease.

Many people think "cancer treatment" when they hear the term *chemotherapy.* Although it is true that chemotherapy is still the primary treatment for cancer, it is also used to control parasitic diseases caused by fungi, **protists**, and worms. A parasite is an organism that lives in or on the living tissue of a host organism at the expense of that host. The first chemotherapy chemical was discovered accidentally during World War II. Military researchers noticed that a toxic chemical agent known as mustard gas killed white blood cells and cancer cells. This led to the treatment of lymphoma, a cancer of certain white blood cells, with mustard gas in the 1940s. The treatment was not fully targeted because the mustard gas also killed healthy cells.

Since the 1940s, pharmaceutical companies have developed a large variety of chemotherapy drugs used in cancer treatment. These drugs are used alone or in combination with other types of drugs or therapeutic practices. A wide variety of chemotherapy drugs make up the anticancer arsenal. Chemotherapy chemicals are categorized according to their effect on the cell chemistry and cell reproduction. Targeted chemotherapies were difficult to develop because most of the past strategies used chemicals that harmed normal cells as well as diseased cells and infectious agents. In 1998, various scientists at the 40th Annual Meeting of the American Society of Hematology (ASH) presented research studies showing how cancer cells and any type of microbes could be killed without harming healthy body cells. This new strategy was made possible because of the findings of basic research studies on the function of the human immune system.

Other types of targeted chemotherapies developed along with the commencement of the **Human Genome Project** in 1990. The project, which was coordinated by the U.S. Department of Energy and the National Institutes of Health, was initiated to sequence all the genetic information in each of the 23 human chromosomes. The Human Genome Project evolved into a legion of scientists interested in identifying all the approximately 20,000–25,000 genes and placing the information in supercomputer databases. It also sparked a great race between government-funded projects and the privately financed research of Dr. Craig Venter. Venter's research challenged the government to hurry the unraveling of the human genome. These databases are fed into software that determines the genomic and metabolomic characteristics of the genes. Scientists can then develop chemotherapies that specifically target certain genes associated with combating cancer and infectious diseases. The targeted drugs interfere with characteristics found predominantly or solely in the cancer cells or infectious agents. This then permits the chemotherapy to abate the disease without harming body cells.

WHAT IS PHARMACOGENETICS?

Pharmacogenetics is the study of how a person's particular genetic makeup affects his or her response to therapeutic treatments. It is a more accurate way of studying the mechanism of action for drugs, which is divided into two categories of investigation—**pharmacodynamics** and **pharmacokinetics**. Pharmacodynamics includes the processes involved in a drug's effect on a cell, tissue, organ, or the whole body. The effectiveness of pharmacodynamics is dependent on a person's genetic makeup. Variations of certain genes influence the pharmacodynamic properties of a treatment. Pharmacokinetics is the study of drug metabolism related to the time required for **absorption**, duration of action, distribution in the body, and removal from the body. These processes are affected by the body fat composition, ethnic origin, gender, size, and weight of a person.[4]

A study published in the October 1984 Proceedings of the National Academy of Sciences[5] was one of many research endeavors of that period that heralded the principle of pharmacogenetics. The study explains why people responded to alcohol in different ways. It was discovered that different genetic types of a chemical called alcohol dehydrogenase affected a person's ability to break down alcohol in the blood. Many studies showed that Africans and Asians were more likely to be affected by lower amounts of alcohol than were people of northern European ancestry. These findings spurred scientists to investigate the genetics of other metabolic pathways that affect the body's ability to handle drugs and toxins.

Another study called "Pharmacogenetics: The Slow, the Rapid, and the Ultrarapid" published in the March 1994 issue of *Proceedings of the National Academy of Sciences* confirmed that variations in certain genes contributed to the different dosages and treatment strategies needed for drug therapies.[6] It appears that a chemical pathway called the **cytochrome P450 system** varied from person to person. This chemical pathway is responsible for the processing of drugs and toxins in the body. The cytochrome P450 systems of some people remove drugs rapidly from the body while the systems of other people degrade the drugs slowly. This explains why particular amounts of a drug are needed to produce an effect in different people. Physicians began using this information to adjust the type of drugs and dosages needed to treat patients. Pharmaceutical companies took the research into consideration when designing and manufacturing drugs. They started making variable dosage units and paid attention to how each component of the drug could interact with the pharmacodynamics of the cytochrome P450 systems of different individuals.

Other variations in drug metabolism pathways determine drug side effects. Side effects might be headaches, nausea, a rise in blood pressure, and liver damage. This has become another important consideration in using pharmacogenetics principles to design targeted drugs. Much care is taken of other chemicals that are mixed with drugs. Most drugs contain additives called **bioavailability enhancers**, **excipients**, **emulsion** and **microemulsion compounds**, **liquid vehicles**, and **lubricants**. Bioavailability enhancers are chemicals that help drugs enter the blood and the cells causing the ailment. Emulsion and microemulsion compounds help drugs form a uniform mixture. Chemicals such as complex carbohydrates and certain types of minerals are used as emulsion and microemulsion compounds. Certain emulsion and microemulsion compounds assist with the time release of a drug and even help the drug pass through the digestive system without being broken down. Liquid vehicles are fluids that are mixed with the drug to make it uniform and palatable. Lubricants help drugs flow through the machinery and equipment used to mix and mold the drugs. All these drug additives can interact with a person's cytochrome P450 system, metabolism, and immune system to influence the effectiveness and safety of the drug. Drug targeting also takes this factor into account when developing precision individualized drugs.

CURRENT AND FUTURE USES OF PHARMACOGENETICS

In 2006, an article in *Pharmacological Reviews* provided a current synopsis of pharmacogenetics.[7] Scientists at the Department of Medicine, Christchurch School of Medicine and Health Sciences in Christchurch, New Zealand, commented:

> The application of pharmacogenetics holds great promise for individualized therapy. However, it has little clinical reality at present, despite many claims. The main problem is that the evidence base supporting genetic testing before therapy is weak. The pharmacology of the drugs subject to inherited variability in metabolism is often complex. Few have simple or single pathways of elimination.

Their unfavorable forecast about the future of pharmacogenetics is not halting the numerous research studies being conducted at pharmaceutical companies and universities. Currently, there are no pharmacogenetic drugs being used as regular clinical treatments. However, potential therapies are under development.

The first step in developing pharmacogenetic drugs involves the identification of the genes that interact with drug action and metabolism. Biochips are being developed to identify clusters of genes associated with particular metabolic pathways (Figure 4.1). One type of biochip will use DNA to detect the presence of genetic variations called **single nucleotide polymorphisms**, or SNPs. SNPs are common mutations consisting of a change at a single base in the DNA. These biochips will detect SNPs that affect drug action and metabolism. Another type of biochip, called a microarray, will monitor gene activity in the presence of particular drugs. A microarray is a biochip for studying how large numbers of genes interact with each other. Various biotechnology companies produce customized microarrays and computerized equipment that reads and interprets the data.[8]

A wealth of publications on possible pharmacogenetic drugs has identified several diseases in which targeted drugs are likely to develop within the next two or three years. Asthma, breast cancer, and colorectal cancer are currently being investigated for treatment using pharmacogenetics drugs. Drugs will be tailored to specifically target the aberrant genes that cause the diseases. A strategy called **RNA interference** works by using specially made RNA strands that inhibit the expression of particular genes. RNA stands for ribonucleic acid, which is a chemical similar to DNA from which proteins are made. The technique of RNA interference uses a strand of RNA to block the action of a disease-causing gene. RNA interference can also be used to stop the growth of microorganisms that cause disease in humans and animals.

Other strategies for turning on and off genes involved in genetic and infectious diseases are also under investigation. Scientists are currently testing chemicals that modify gene function to treat AIDS, allergies, bacterial infections, depression, cancer, chronic pain, heart disease, and seizures. Many of these chemicals are artificial compounds created in laboratories. However, scientists are also relying on natural pharmacogenetic drugs produced by bacteria, fungi, and plants. Researchers at the College of Pharmacy of

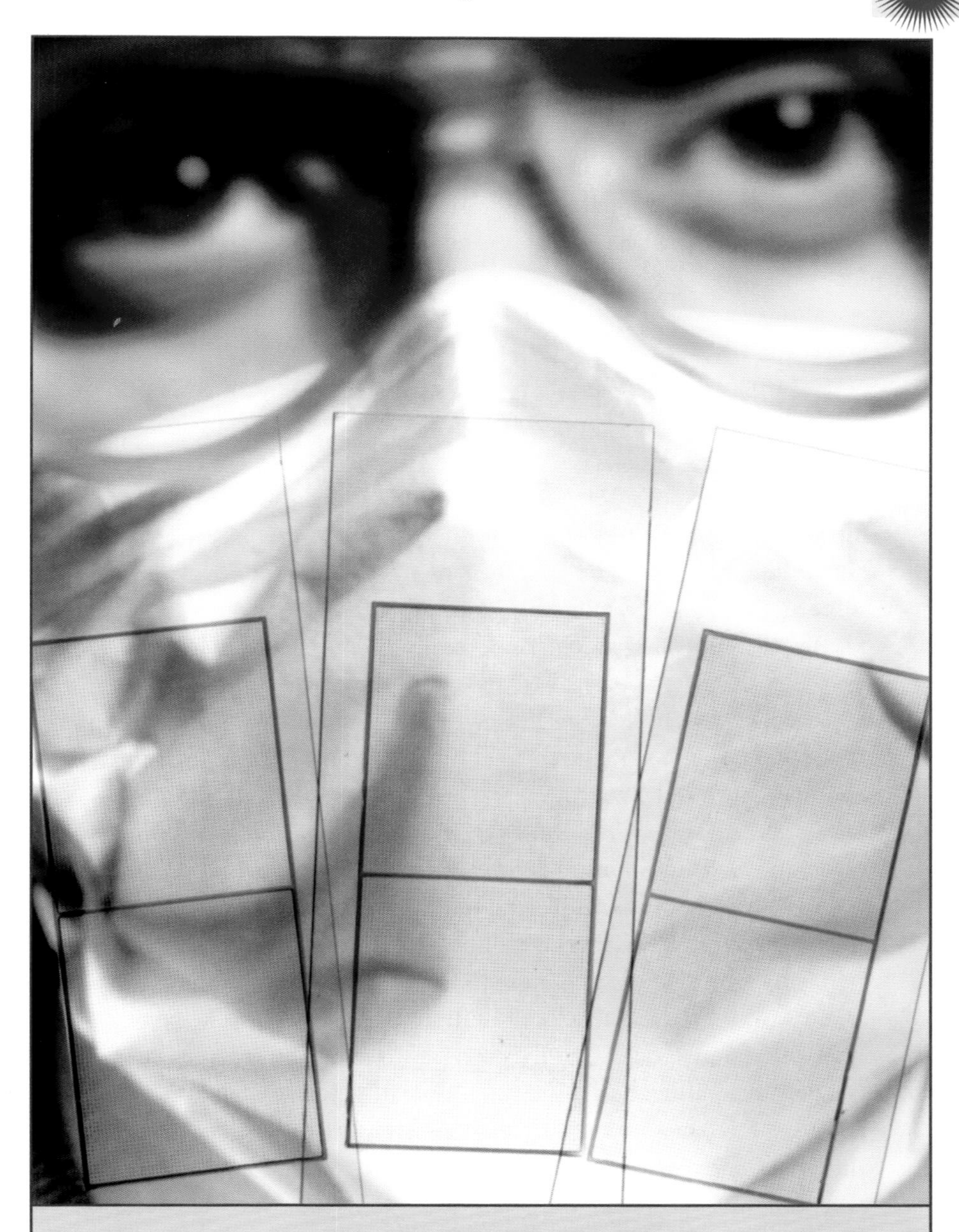

Figure 4.1 Devices called biochips identify genes that help with the development of pharmacogenetic drugs. Above, a technician holds up three biochips. A single biochip can conduct thousands of chemical and biological reactions in just minutes.

the University of Texas at Austin are currently testing a drug called venlafaxine in pharmacogenetic trials. Venlafaxine, or Effexor, is a new antidepressant that does not work the way currently used antidepressants do. It interacts with nerve cells in a more specific manner than traditional antidepressant drugs. Therefore, it works more accurately with fewer side effects.[9]

WHAT ARE PRECISION VACCINES?

Another area of targeted medicine is precision vaccines. Vaccines are the drugs of a medical procedure known as **immunization**. Immunization is a process by which protection from an infectious disease or cancer is administered. Any treatment using a vaccine is called vaccination. The idea of vaccination was formally recognized as a medical treatment in England in 1792. English author Lady Mary Wortley Montagu tried a method of treating smallpox that she learned on her travels to Turkey. She discovered that people in Turkey infected themselves with a mild form of smallpox as a way of warding off the debilitating form of the viral disease. They did this by collecting pus from smallpox sores and scratching it into the skin of healthy individuals. Unfortunately, the strategy did not work well in England and possibly helped spread the disease.

In 1792, an English physician named Edward Jenner modified Montagu's approach. He noticed that people who worked with cattle that had a disease called cowpox were immune to smallpox. So he collected the cowpox virus from cattle and scratched it into the skin of healthy people. The technique worked, and ultimately it became an accepted method of preventing smallpox. This strategy became known as vaccination because the Latin term for *cow* is "vacca." Vaccination became a standard lifesaving medical procedure. The discovery of smallpox vaccination has eradicated the disease that killed 500 million people worldwide during the twentieth century.[10]

Contemporary vaccines come in two forms: active vaccines and passive vaccines. Active vaccines work by stimulating the immune system to fight an infection. They provide a long-lasting immunity that may persist for 5 to 10 years. Active vaccination is commonly given for diseases such as influenza and tetanus. Passive vaccines supplement the immune system. They are temporary and must be administered with each exposure to a disease. Passive vaccines are usually given to fight off a variety of highly contagious diseases that do not respond well to active vaccines. It is generally administered for viral diseases such as chicken pox, German measles, hepatitis A, and measles.

Active vaccines are made in several ways. The first active vaccines were produced using weakened or killed disease organisms. In 1952, Jonas Salk developed a vaccine for polio this way at the University of Pittsburgh. Polio, or poliomyelitis, is a viral disease that causes inflammation of nerve cells in the brain stem and spinal cord. It usually results in loss of muscle control. In some cases people had to be placed in machines called iron lungs that helped them breathe after they lost control of their respiratory muscles. Salk used a mixture of the three types of virus, grown in monkey kidney cells grown in culture. He used a chemical preservative called formaldehyde to weaken the virus. Active vaccines are somewhat targeted against the disease-causing organism.

Passive vaccines are usually mixtures of proteins called antibodies or **immunoglobulins**. Antibodies are immune system chemicals that attach to foreign substances in the body. Immunoglobulins are proteins produced by the white blood cells that help battle foreign substances in the body. The immunoglobulins developed in the laboratory are obtained from laboratory animals or cell cultures. They act like antibodies and signal the body to remove disease-causing organisms. Immunoglobulins are not highly targeted and produce a minimum immune response against the organisms causing disease. The immunoglobulins do their job much like antibodies by adhering to invading organisms in the blood or in body tissues. White blood cells then attack any organism sticking to the immunoglobulin proteins.

Active and passive vaccination is effective in preventing many types of diseases. Vaccination is one of the most effective and favorable strategies for protecting people and animals against infectious diseases. Vaccination programs worldwide have eradicated smallpox, nearly eliminated polio, and greatly reduced the incidence of common diseases such as chicken pox, diphtheria, German measles, influenza, measles, mumps, pertussis, and tetanus. Unfortunately, the benefits of vaccination are sometimes outweighed by their limited effectiveness against many diseases. Certain disease organisms change so quickly that they cannot be controlled by vaccines from one year to the next. Also, no vaccine is perfectly safe and may lead to serious side effects in some instances.

It is hoped that the development of precision vaccines will remove the dangers and ineffectiveness of traditional vaccines. Effective precision vaccines specifically target the invasive organism or cancer cell using highly specific artificial antibodies. The highly specific feature of the **antibody** is called a **paratope**. The paratope is the region of an antibody that attaches to a chemical. This attachment is based on the attraction of the paratope to a particular shape and chemistry of the chemical. The region of the chemical that binds to the paratope is called the epitope. Scientists are currently learning how to engineer paratopes that bind to epitopes in a more effective and selective manner. Current vaccines are only about 60–80% effective at eliciting the necessary immune response. Anticancer vaccines are much less effective.[11]

Precision vaccines will also have to better assist the immune system with killing or removing infectious agents than traditional vaccines. Scientists are currently modifying the chemistry of vaccine antigens so that they act as stronger immune system signals or directly kill a targeted organism. Traditional vaccines merely identify the foreign substance so that the immune system can use its regular mechanisms to destroy or remove the material. Precision vaccines will be able to enhance the immune response or degrade the foreign material by attaching molecules called **ligands**. A ligand is a chemical attached to another to impart specific characteristics.

CURRENT AND FUTURE USES OF PRECISION VACCINES

Currently, there are no precision vaccines in regular use for treating disease. However, there are many efforts to develop these vaccines. Researchers are directing their efforts on precision vaccines against cancers and diseases caused by fungi, protista, and worms. Early efforts with traditional vaccines to cure these diseases have failed or have proven ineffective. These vaccines did not work well because antibodies used to target the cancer cells and organisms were not able to produce a large enough immune response to completely eradicate the disease. This permitted the cancer and organisms to develop defense mechanisms for avoiding the vaccine.

The breakthrough strategy for precision vaccines uses an antibody connected to a chemotherapy molecule. In 1998, a team of scientists from the Fred Hutchinson Cancer Research Center collaborated with researchers from 13 leading leukemia centers including MD Anderson Cancer Center in Texas, the University of Chicago Medical Center, and the University of Pennsylvania Cancer Center to produce the first precision vaccine. It was composed of a targeted antibody that delivers chemotherapy drugs to cancer cells. The treatment developed by the research team was called **conjugate therapy**. Conjugate therapy is treatment with a drug composed of two different therapeutic agents attached to each other.[12]

The precision vaccine they developed was composed of an artificial antibody that selectively binds to leukemia cells. Leukemia is a cancer of white blood cells. Attached to the antibody was a chemotherapy agent called calicheamicin. Calicheamicin is a potent toxin that is effective against leukemia. This conjugant therapy was tested on patients with acute myelogenous leukemia. Initial findings showed that the precision vaccine was much more effective than using calicheamicin, which was only 40% effective at treating acute myelogenous leukemia. In addition, the side effects of the conjugated therapy were mild compared to the traditional chemotherapy. Precision vaccines related to this are being investigated to stem the spread of **bird flu**. Bird flu, or avian influenza, is

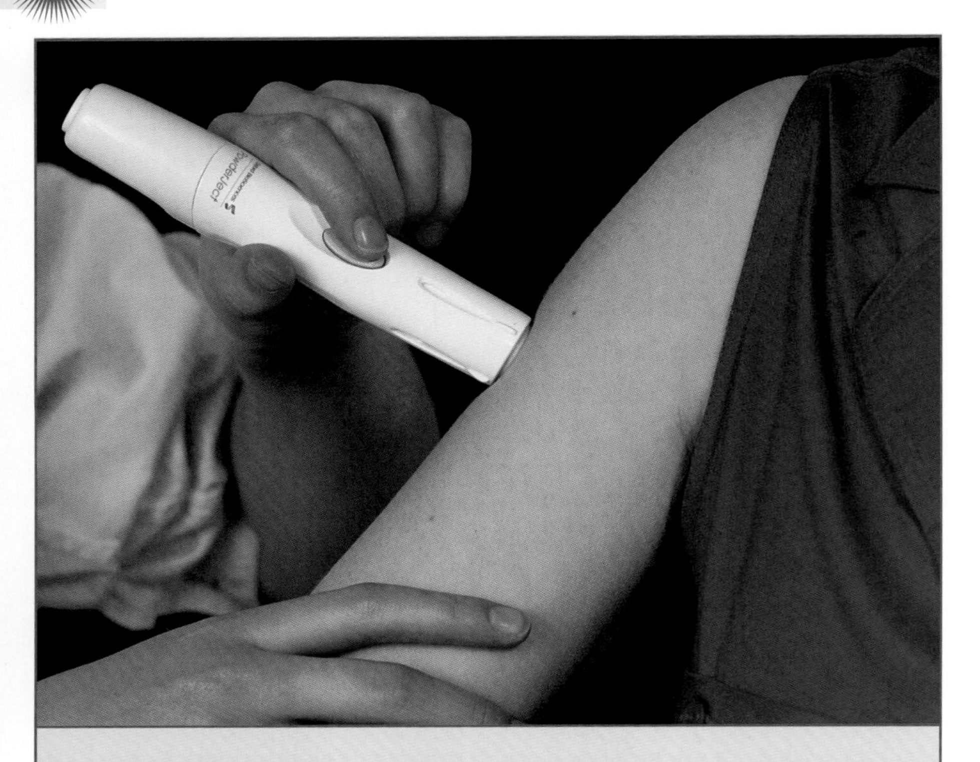

Figure 4.2 In the photograph above, a PowderJect needle-free injection system is pressed against an arm. A needle-free injection causes less pain and leaves a smaller wound on the patient.

a type of virus that harms the respiratory system of birds and can be spread to other animals.

A 2002 study carried out at the Department of Immunohematology and Blood Transfusion, Leiden University Medical Center tested another precision vaccine strategy. The research team developed a conjugated antibody with a chemical that attracts immune system cells called **cytolytic T lymphocytes**. Cytolytic T lymphocytes are white blood cells that secrete toxins to combat cancer and parasite infections. Initial studies on rats showed that the treatment is highly effective against certain cancers. It is being targeted against human papillomavirus, which causes cervical cancer. The researchers hope to use the vaccine as a **preventative medicine** measure against other cancers, parasitic infections, and sexually transmitted diseases. This vaccine will boost the immune system so that exposure to the disease induces a rapid and effective immune response that completely removes the disease agent before it causes illness.

Precision vaccines are benefiting from better vaccine delivery technologies. Scientists have developed new types of delivery systems, such as air-powered injectors (Figure 4.2), that do not require needles to get the vaccine into the blood or body tissues. Air-powered injectors are less painful than a needle and leave a smaller wound that is more likely to heal quickly. Another delivery strategy uses patches that permit the precision vaccine to absorb into the skin. These will resemble the **transdermal** patches currently used in birth control and smoking-reduction patches. Oral precision vaccines are also on the horizon. These vaccines can be administered through a special pill that protects the vaccine from being digested in the stomach and small intestine. Some researchers are hoping to produce crop plants that contain the precision vaccine. This will provide a simple means of introducing the vaccine rapidly into large populations of people exposed to an infectious disease.[13]

REFERENCES

1. B. Korber, M. LaBute, and K. Yusim. "Immunoinformatics Comes of Age." *LoS Computational Biology.* 2(6) (2006): 71.
2. Novartis Foundation. *Immunoinformatics: Bioinformatic Strategies for Better Understanding of Immune Function.* Hoboken, N.J.: John Wiley and Sons, 2003.
3. L.M. Hernandez and D.G. Blazer, eds. *Genes, Behavior, and the Social Environment: Moving Beyond the Nature/Nurture Debate.* Washington, D.C.: National Academy Press, 2006.
4. S. Harbron and R. Rapley. *Molecular Analysis and Genome Discovery.* Hoboken, N.J.: John Wiley and Sons, 2004.
5. R. Bühler, J. Hempel, R. Kaiser, J.P. von Wartburg, B.L. Vallee, and H. Jörnvall. "Human Alcohol Dehydrogenase: Structural Differences Between the Beta and Gamma Subunits Suggest Parallel Duplications in Isoenzyme Evolution and Predominant Expression of Separate Gene Descendants in Livers of Different Mammals." *Proceedings of the National Academies of Science.* 81(20) (1984): 6320–6324.
6. U.A. Meyer. "Pharmacogenetics: the Slow, the Rapid, and the Ultrarapid." *Proceedings of the National Academies of Science.* 91(6) (1994): 1983–1984.
7. Sharon J. Gardiner and Evan J. Begg. "Pharmacogenetics, Drug-Metabolizing Enzymes, and Clinical Practice." *Pharmacol Reviews.* 58 (2006): 488-520.
8. I.P. Hall and M. Pirmohamed, eds. *Pharmacogenetics.* Oxford: Informa Healthcare, 2006.
9. National Research Council. *Advancing the Nation's Health Needs: NIH Research Training Programs.* Washington, D.C.: National Academy Press, 2005.
10. M. Kennedy. *A Brief History of Disease, Science and Medicine.* Mission Viejo, Calif.: Asklepiad Press, 2004.
11. B.R. Bloom and P.H. Lambert. *The Vaccine Book.* Oxford: Academic Press, 2003.

12. K.R. Stratton, J.S. Durch, and R.S. Lawrence, eds. *Vaccines for the 21st Century: A Tool for Decisionmaking.* Washington, D.C.: National Academy Press, 2000.

13. B. Ludewig and M.W. Hoffmann, eds. *Methods in Molecular Medicine: Adoptive Immunotherapy Methods and Protocols.* Totowa N.J.: Humana Press, 2005.

FURTHER READING

Books

Borem, A., F.R. Santos, and D.E. Bowen. *Understanding Biotechnology.* San Francisco: Prentice Hall, 2003.

Gallagher, W. *I.D.: How Heredity and Experience Make You Who You Are.* New York: Random House, 1996.

Shmaefsky, B.R. *Applied Anatomy and Physiology: A Case Study Approach.* St. Paul, Minn.: EMC/Paradigm, 2007.

Web Sites

BBC News—Health

http://news.bbc.co.uk/1/hi/health/default.stm

Biotechnology Industries Organization

http://www.bio.org

National Center for Biotechnology Information

http://www.ncbi.nlm.nih.gov

National Library of Medicine

http://www.nlm.nih.gov

Medical Instrumentation

WHAT'S IN THE NEWS?

In 2006, systems design engineer John Yeow at the University of Waterloo in Canada presented his vision of a new era for medical instrumentation. His research on nanotechnology holds promise for innovative medical instruments that permit more effective medical diagnosis and treatment. He claims that many of the contemporary instruments are based on traditional designs that date from decades if not thousands of years ago. According to numerous physicians and health professionals, many of the instruments used in delicate surgical procedures are clumsy to use. Even a simple scalpel requires critical skills to prevent cutting too deeply or too shallowly. Plus, the injuries caused by traditional scalpel cuts do not always heal properly and sometimes leave scars and blood vessel damage.[1]

Yeow is primarily concerned with improving medical instruments used for diagnosing diseases. He and other researchers throughout the world are using nanotechnology principles to shrink the size of large diagnostic instruments that must be inserted into the body. One instrument that needs to be reduced in size is the **endoscope** (Figure 5.1). An endoscope is a tubular medical diagnostic instrument that is inserted into the body to view its internal components. The endoscope is usually passed through a natural opening such as the mouth, rectum, or vagina. Some endoscopes are inserted into small surgical incisions for investigating features such as joints that lie under skin and muscle. Any medical procedure using an endoscope is called an endoscopy.

A typical endoscope inserted into the body is a somewhat flexible tube that can exceed 1.5 cm (0.6 inches) in diameter. That might seem like a thin diameter, but a tube with that thickness is not very flexible and is very large compared to the throat and anus. A typical endoscope also has a large eyepiece for viewing by the physician's eye or by a video camera. A lamp installed within the tip of the endoscope is needed to illuminate the dark body cavities. The typical endoscopic procedure is very uncomfortable to people

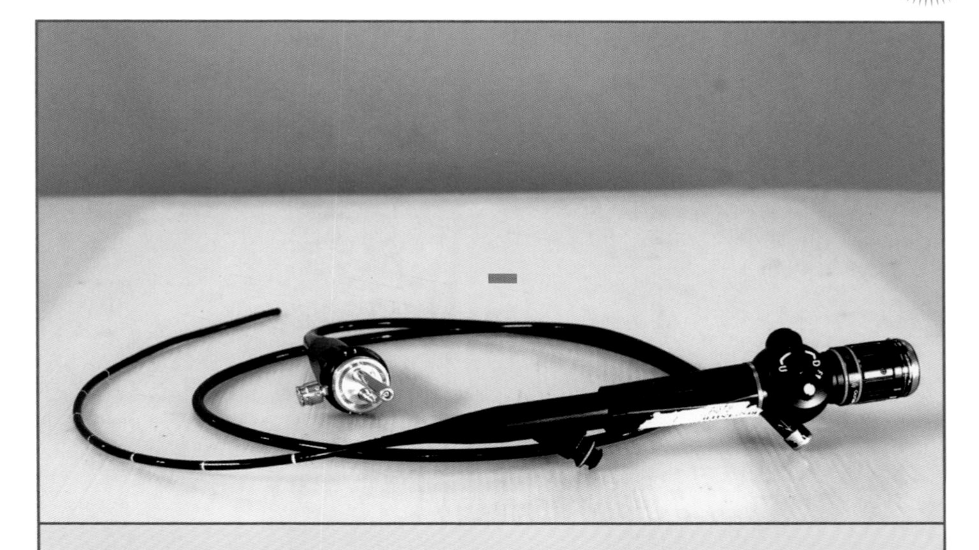

Figure 5.1 Traditional endoscopes are cumbersome to use and painful to the patient. Newer endoscopes are being designed to remedy these problems.

and many endoscopes scratch the lining of the throat, rectum, and reproductive tract. It is not unusual for a person to need medical treatment to alleviate irritation and infection caused by traditional endoscopes inserted into body openings.[2]

Yeow is hoping to develop a micromachine-based endoscope as thin as sewing thread. It is expected to be 100 to 1,000 times narrower than the traditional endoscope. These narrow endoscopes will be very flexible and cause almost no discomfort to the patients. In addition, physicians will have an easier time inserting the scope into patients with diseases that restrict the body openings. These new endoscopes will be able to diagnose and treat diseases at the same time. Special circuits attached to miniature sensors will eliminate the need for lights and cameras. The images could be sent to a computer that provides real-time, color-enhanced videos and still pictures. Small surgical instruments built into the endoscope can be used to remove tumors or seal damaged tissues.

Figure 5.2 A new high-tech endoscope is displayed above. The endoscope, with a diameter of only 5 millimeters (0.2 inches), will be able to move throughout the gastrointestinal tract, scanning the area, and delivering drugs to the affected region. In the future, nanotech engineers hope to design an endoscope that is a fraction of the size of this one.

Many nanotechnology engineers are hoping to see the same miniaturized technology applied to endoscopes used to view diseases under the skin (Figure 5.2). It may be possible to design endoscopes for this purpose that are the thickness of a pin. Many researchers are hoping that endoscopes will be replaced by other technologies. Yeow already foresees the development of micromachines that travel within a patient's body while analyzing and even correcting diseased tissues.

INTRODUCTION

Early medical instruments were cutting tools designed to do minor surgical treatments that may or may not have a valid therapeutic basis.[3] In ancient times, sharpened rocks were used to cut skin and drill holes in bones. Ancient Egyptian physicians went through years of training at temple schools to learn a variety of medical practices, many of which used surgical instruments. The earliest practices involved cleaning and bandaging wounds. Materials such as moss and leaves were used to make bandages. Medicines consisted of plant extracts. It is likely that the first surgical procedures included **trepanation,** which entailed using pointed stones to drill holes into the brain. Trepanation was believed to alleviate pain caused by head injuries. Eventually, iron and steel surgical instruments were fashioned for performing surgery to repair internal wounds and even perform **Caesarian deliveries**.

The basis of traditional modern surgical tools dates from the fifth century B.C. Roman medical instruments. Early scalpels were used to make incisions in skin and muscle. Bone levers were used to move fractured bones into position and were possibly used for pulling out teeth. Large forceps were used to remove fractured bone fragments and might have served as devices for gripping a baby's head during childbirth. A wide array of cupping vessels was used to drain body fluids. Small cupping vessels were used to draw blood from arms and lower legs. Large cupping vessels had been used for bigger body parts such as the back, belly, and thighs. Adhesion and contraction tubes were used to reduce improper healing of nose, rectum, and vaginal operations. They were also inserted into body parts to help drain fluids.

Ancient Greek and Roman instruments were used for investigating the digestive system and urinary tract. Curettes, forceps, probes, and speculums were used to assist with examinations of the rectum and vagina. Even catheter tubes were available for draining the urinary bladder. Very few medical instruments were developed between ancient times and the 1800s. Then, in 1816, the first stethoscope was invented by a French physician named René Laennec. It was composed of stacked paper rolled into a solid cylinder very much resembling the cardboard making up the center of a toilet paper roll. Eventually wood replaced the paper roll. This stethoscope was inconvenient to use because the physician had to hold his or her head almost directly on the chest. In 1902, the modern combination, or bell stethoscope, was developed. It was more comfortable to use and permitted the physician to hear heart sounds much better. The invention of the stethoscope facilitated the development of the first blood pressure cuff or sphygmomanometer by Samuel von Basch in 1891.

The repertoire of medical instruments was somewhat limited until the beginning of the 1900s when the first electrical circuits suitable for electrical instruments were in common use. Until then, electricity was primarily used to run lightbulbs, simple radios, telegraphs, and telephones. Microscopes were greatly improved using electrical lighting sources. This made them important tools for studying disease by examining blood, tissues, and urine. The first accurate electrocardiogram (ECG) instrument was invented by Willem Einthoven in 1901—but it weighed 600 pounds (272 kilograms). However, it served as the prototype for the modern ECG, which led to the development of the **electromyogram** in 1934 and the electroencephalogram in 1936. Electromyograms (EMG) evaluate nerve and muscle function. The medical instruments developed in the 1950s permitted physicians to carry out delicate surgical procedures not possible with older instruments. In addition, new instruments for diagnosing disease were being developed.

WHAT ARE MEDICAL INSTRUMENTS?

There are three major categories of medical instruments: delivery, diagnostic, and surgical. Medical delivery systems are instruments that transfer medications to specific body parts. Diagnostic systems assist with the identification of disease. Surgical instruments are tools for repairing diseased and injured body components, including devices used in dentistry and ophthalmology. Each of these categories of instruments is divided into subcategories based on specialized uses. The design of medical instruments requires the cooperative efforts of engineers, physicians, and scientists who contribute their expertise to ensure the usability and safety of the instruments. In addition, medical instruments must meet strict safety government guidelines. Almost every country has requirements for instrument design. In addition, there are international guidelines that ensure the quality of medical instruments sold throughout the world.[4]

The subcategories of medical delivery systems are infusion, needleless (without needles) injection, and syringe injection. Infusion systems slowly pump chemicals into the bloodstream or a body cavity. They can be used as stand-alone devices or can be attached to **intravenous** bag setups. Some delivery systems are designed to rapidly introduce medications in the body. In many cases it is essential to place medications slowly into the body. Devices that accomplish this are called sustained-release delivery systems. Delivery systems can be automatic or manual. Automatic systems are designed to introduce the medication at particular intervals. Manual systems are controlled by the patient or a medical professional.[5]

Traditional infusion systems are placed outside of the body and pass along the fluid into the body through a needle inserted into the skin. The typical system is operated by an electrical pump that pushes the fluid through a narrow tube. There are different types of infusion pumps. A piston pump sucks fluid into a chamber and then pushes the fluid through a tube. Peristaltic pumps use rollers that rub against the tube and squeeze the fluid through the tube. Some infusion systems are attached to tubes that pass through surgical openings cut into a body cavity. Infusion can also be introduced into the body through tubes that pass into the mouth, nose, or rectum. Infusion systems are generally designed to deliver liquid forms of medicine.

Needleless injection systems include a broad category of delivery systems in which liquid or powder medications are introduced into the body fluids without puncturing the skin with a syringe needle. Nebulizers are needleless medical devices that deliver liquid or powdered medication in the form of a mist to the airways. They can be used to deliver many types of medications, including antibiotics and certain types of vaccines. They have the same effect as a needle injection in that they deliver drugs quickly into the bloodstream. The drug rapidly enters the blood after dissolving into the respiratory membranes.

Some delivery systems overlap in function and can be used for infusion or needeless injection. Transdermal systems are called passive delivery systems because they do not use a needle or pumping device. They use a principle called absorption and **diffusion** to introduce the medicine into the body. Absorption is the process of a medication passing through the skin. Diffusion is the natural movement of particles. Transdermal patches are complicated devices that ensure accurate delivery of the chemical into the body. A drug reservoir holds the medicine in a solution that helps carry the medicine across the skin.

The flow of medicine through a transdermal patch is controlled by a special material called the drug-release membrane (Figure 5.3). These membranes must be tailored to match the size and particular chemical characteristics of the medication in the drug reservoir. Special adhesives are used to attach the patch to the skin. The adhesive is designed to be removable when needed. However, it cannot come loose from the skin when exposed to sweat or soapy water. In addition, the adhesive should not cause skin irritation.

Syringe injection systems have been used to deliver liquid medicines since 1670. The first syringe with a needle small enough

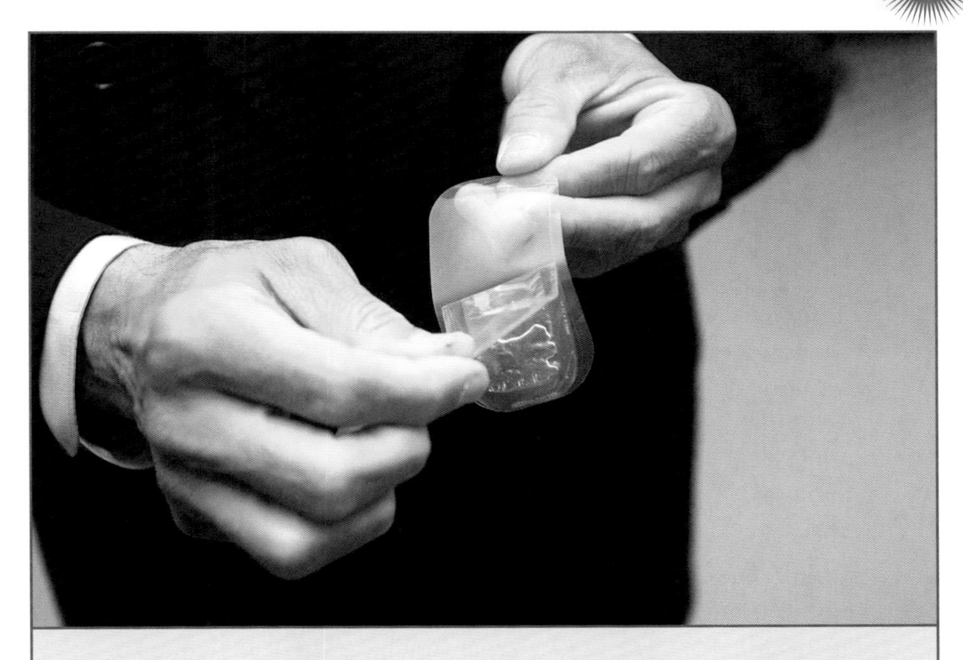

Figure 5.3 Transdermal patches are passive infusion devices that deliver precise amounts of a medication into the body.

to inject medicine into a blood vessel was developed by Charles Gabriel Pravaz and Alexander Wood in 1853. Syringes are composed of a housing, plunger, and needle. The housing is calibrated to hold a specific volume of medicine. Most syringe housings have markings that tell how much solution is contained or being delivered. The plunger is designed to push the liquid out of the housing and into the needle. Plungers can be manual or mechanized. Manual syringes are the most commonly used syringe for drawing out body fluids and delivering drugs. The speed and amount of delivery is controlled using hand motions. Mechanized syringes have plungers operated by an electric pump. Syringe needles come in a variety of diameters. The selection of a particular needle is determined by the type of injection being done and by the amount of solution being delivered.

Diagnostic Medical Instruments

Diagnostic medical instruments are categorized according to the body measurements being collected and analyzed. **Electrophysiology** instrumentation includes instruments that monitor the electrical activity of the brain, heart, muscles, and nerve tracts. The term *electrophysiology* refers to measuring the electrical activity of body components. Some instruments, such as ECG, EEG, and EMG, amplify electrical impulses from the body. The impulses are compared to a database of normal and diseased readings. Traditional electrophysiology instruments recorded data on a chart that was then interpreted by a physician. New instruments feed the measurements into a computer that helps analyze the patient's condition.

Imaging technologies form another category of diagnostic medical instrumentation. Modern medical imaging is grouped into **magnetic resonance imaging**, microscopy, radiography, and ultrasound. Magnetic resonance imagining (MRI) is a complex technology that uses powerful magnets to measure chemical characteristics of tissues. The magnets cause chemicals in the body to interfere with radio waves generated by the instrument. Each chemical in the body gives off a unique interference signature that characterizes the molecule. The signature is then interpreted by a detector that builds a chemical map of the body. This map appears as a colored image of the body. A device called a scanner, focused on small points inside the patient's body, characterizes the type of tissue based on the unique chemical features. The instrument is sensitive enough to detect differences in diseased and normal tissues. MRI can identify subtle disease conditions that are not detectable with other types of medical testing.

Medical microscopy is an imaging technology that uses light microscopes and electron microscopes to visually and chemically analyze body fluids and cells. Traditional light and electron microscopes provide details about cell structure. The cells are typically colored with stains that identify particular structural features of the cells and tissues. Images are generally recorded as digital images that can be analyzed using special software that detects disease. New microscopes, such as fluorescent microscopes, permit the physician to investigate cell functions. Certain dyes that fluoresce, or glow, in the presence of certain chemical reactions can indicate the activity of DNA or metabolic reactions.

Radiography or X-ray imaging is the oldest of the medical imaging technologies. It was first used in medicine in 1986 only one year after X-rays were discovered by Wilhelm Conrad Roentgen. It was first used to observe injury to bones. The first X-ray machines could only examine small body parts such as the extremities and the head. Larger modern machines permit the whole torso to be X-rayed. Earlier machines had unshielded X-ray tubes that irradiated the whole room with the dangerous radiation. Modern shielded tubes focus the X-rays on the body parts being examined. Advances in X-ray film make it possible to see fine details on even minute body parts.

Improvements in electrical circuitry permit physicians to adjust the X-rays for examining subtle differences in tissue density. Fluoroscopy is a type of radiography in which X-rays are used with special dyes to produce live, moving pictures on a screen. Physicians can add dyes to the blood for investigating blood flow through various body parts. This a common practice for visualizing the health of blood flow in the heart. Dyes can also be introduced into the digestive system to look for abnormalities such as growths and ulcers. Computed **tomography** (CT), which is sometimes called a CAT scan, makes use of special X-ray equipment to produce detailed cross-section sliced images of the body. Newer CT instruments can produce three-dimensional images. Another type of tomography called **positron** emission tomography computed tomography (PET/CT) uses positively charged subatomic particles called positrons in place of X-rays. A PET/CT instrument has a more open design than traditional CT and does not enclose the patient.

Ultrasound is an imaging technology that works by the same principle as radar. *Radar* is an acronym for *radio detection and ranging*. It bounces radio waves off an object as a means of detecting size, location, and position. The reflected waves, or echoes, are converted into an image that shows internal structures. Ultrasound is commonly used to view the development of a baby within its mother. The sound waves are not destructive to the body, as X-rays are. Doppler is a type of ultrasound technology generally used to monitor the flow of blood through blood vessels and the heart. It can detect unusual patterns that indicate various types of cardiovascular disorders. Earlier ultrasound instruments were very large. Improvements in detectors and electrical circuitry reduced the size of the units so that they can now be carried in a briefcase.

Clinical chemistry testing instruments include a wide array of diagnostic machines that characterize body fluids and tissue samples for disease and injury. Blood analyzers are the most commonly used clinical testing instruments. The heart of these analyzers is an instrument called a **spectrophotometer**. A spectrophotometer is an instrument that measures the amount of light reflected from a specimen when illuminated by a controlled light source. Different chemical components of blood can be analyzed using the spectrophotometer. Various types of blood analyzers are used during a blood examination to determine the quantities and types of carbohydrates, fats, hormones, metabolic wastes, and proteins in blood. Blood gas analyzers are designed to detect carbon dioxide, oxygen, and other atmospheric gases in the blood.

Urine analyzers work on the same principle as blood analyzers. However, they differ from blood analyzers in that urine analyzers are calibrated to detect different types of chemical components. Other types of analyzers have been developed to examine spinal fluid and tissue exudates. Exudates are fluids produced by injured tissues. Cytometers are spectrophotometer-based diagnostic instruments that evaluate cells. Most cytometers compute the number and types of cells in a sample. They are very important in determining diseases of the blood. Specialized cytometers can indicate the presence of microorganisms and cancerous cells. Dyes can be added to indicate genetic and metabolic differences between cells passing through the cytometer. Diagnostic **chromatography** systems separate and analyze mixtures of molecules in solution or in a body fluid. They are commonly used to look for hormones, metabolic wastes, and toxins.

Surgical Instruments

The most commonly used surgical instruments are precision tools designed for procedures used during an operation. For the most part, surgical instruments are designed to be extensions of the hand. Surgical instruments are categorized in two ways. They can be classified according to their use in a specialized procedure or their use in a particular body region. Diagnostic surgical instruments are devices such as neurohammers and percussion hammers for testing nerve reflexes. Other diagnostic surgical instruments called calipers assist with the measurement of body parts. Instruments called probes and speculums help the physician look into a body opening.

General surgical procedures use instruments called retractors that help separate the bones, organs, and skin when a procedure is carried out through an open incision. Microsurgery has its own characteristic instruments for making minute cuts and for clamping off tiny blood vessels. Physicians use these instruments while looking through magnifying glasses or microscopes. Plastic surgical instruments are specially designed to carry out cosmetic and reconstructive surgical procedures. Liposuction instruments facilitate the unique tasks needed to remove fat cells from particular body regions. Body region categories include instruments needed for specialized producers on the brain, chest cavity, ears, eyes, genitals, intestines, and teeth.

Most surgical instruments are handled the same way as commonly used hammers, knives, scissors, and wrenches found around the house. Certain instruments such as bone cutters now are electrically operated to increase a physician's speed at completing the task while also reducing fatigue. Most modern surgical instruments are ergonomically designed. **Ergonomics** means that an instrument is shaped to be held using the correct posture and positioning of the body to improve ease of use. Many surgeries can take hours to perform. So, it is important that the physician does not tire or develop muscle cramps while performing the surgery.

FUTURE TYPES OF MEDICAL INSTRUMENTS

Most major improvements in medical instruments have taken place since the late 1980s. These advances were driven by a better understanding of the genetics and **molecular biology** of disease. Medical practitioners now need instruments that meet the demands of the increasingly detailed and precise information about disease diagnosis and treatment. In addition, insurance companies are requiring physicians to provide medical services more accurately and effectively at a minimum cost. Governmental safety guidelines promote medical instrument improvements that make the devices safer to use for both the patients and the physicians. Portability of medical services is another driving force pushing the evolution of medical instruments. It is becoming more common worldwide to bring high-tech medical services to remote locations and populations underserved by major medical centers.

Advancements in drug delivery are leading to the development of "smart" devices that replace some of the medical practitioner's role in delivering medicine. Smart intravenous infusion devices are already in use. They use a calibrated smart-pump technology that delivers a precise amount of medication over a continuous period. Initial trials on elderly patients show a 73% reduction in infusion errors. Many of these errors could have led to death of the patient. The Harvard syringe uses a pumping device that reduces injection errors by pumping a precise amount of medication through the needle. It is possible to use a similar pump on nebulizers and needless injection systems. Added precision is provided by using ultrasound mechanisms to push the mist out of a nebulizer. The only limitation is that the pumps and syringes require a standardized concentration of the drug. This means that drug manufacturers have to develop medications specifically for use in these devices.

Pill dispensing is also being automated to avoid medical errors. Smart pill cases have built-in timers and communication systems that instruct the medical practitioner or patient when to take a pill. One type of dispenser called Monitored Automatic Pill

Dispenser, or MD.2, has a voice alarm. The company that makes the MD.2, e-pill Medication Reminders, also manufactures watches, alarms, and other devices that alert patients and medical practitioners when to dispense medications. Another device called the MedReady medication dispenser has a timer and alarm system that can be set for several days of dispensing. It is now even possible to use an injection system to deliver powdered drugs into the body. A product called PowderJect is a needleless powder injection drug delivery that may one day replace the need to take multiple medications orally.

Internal delivery systems are becoming more popular ways of dispensing hormones and medications that regulate vital body functions. These medical machines are called implantable delivery devices. They deliver small amounts of drugs into the blood or within the fluids surrounding a body organ. The precision engineered machines are composed of a dosing reservoir, microcomputer circuit, pump, and valve. The dosing reservoir holds the drug. The microcomputer circuit, pump, and valve control and deliver the drug. Traditional dosing reservoirs have an external refill port through which medication is added using a syringe. Modern dosing reservoirs that hold living cells are being developed. They will not need to be refilled and will work very much like a natural organ. Biology sensors that help adjust the dose based on the body's needs are also under investigation.

Because home care is becoming a growing part of contemporary medicine, it is not always possible for physicians to keep track of medication delivery. This change is driving an industry that specializes in medical remote monitoring. Smart systems are being designed to communicate dispensing and overdose warning signals to cell phones or recording devices. This information can then alert a family member or medical practitioner about medication compliance by a patient. Remote monitoring can also be used to alert physicians about the proper operation of implanted medical devices. Medical practitioners will be able to use remote systems to deliver drugs. They will be able to send a signal to the delivery system through a cell phone or other two-way communication device. Many of these remote medical devices are based on technology developed by the National Aeronautics and Space Administration (NASA) for use on astronauts.

Diagnostic instruments of the future will be smaller and more accurate than the traditional technology. Neural network computer software may one day replace the physician's role in diagnosing many diseases. This software makes similar decisions to those made by humans and even learns from its experience with various diseases. Initial trials show a high degree of accuracy in diagnosing many types of common diseases. High-tech digital cameras integrated into medical microscopes provide images that can be analyzed by sophisticated diagnostic software. In certain tests, software that analyzes tissues collected in biopsies was able to detect abnormal cells with the same accuracy as that of technicians. A **biopsy** is the removal and analysis of living cells taken from the body. Miniature digital cameras and analytical probes are being placed in endoscopes for recording various types of data. The cameras can record abnormalities that could sometimes be missed during a regular visual examination by the physician. Analytical probes can use **infrared light** to detect subtle abnormalities due to inflammation. Ultrasound or ultrasonic cameras are probes that collect data about blood flow through microscopic blood vessels called **capillaries**. These cameras can also produce chemical signatures of tissues.

Even a routine visit to the eye doctor, or optometrist, will bring a person in contact with increasingly high-tech diagnostic tools. Electronic indentation **tonometry** is being used more often to detect pressure buildup in the eye indicative of the disease glaucoma. Tonometry is a way of measuring the pressure or swelling of a body part. The traditional method of eye tonometry uses a jet of air and visual examination to determine the disease. This new way of doing tonometry is very accurate and may even be able to predict glaucoma even before the disease develops. Tonometry technology is also being used for highly accurate blood pressure monitors that can be placed on a finger or on a blood vessel during surgery. Microscopic tonometry units can even be placed in the body to continuously monitor blood pressure. Infrared detectors and miniature spectrophotometers can be attached to tonometry instruments for monitoring blood chemistry.

Onsite or portable medical diagnostic instruments are becoming a necessity for emergency medical care. It is now almost possible to fit the contents of a medical diagnostic laboratory and imaging facility into a small ambulance or in one room of a tiny clinic. John Yeow at the University of Waterloo is developing the technology to miniaturize hospital X-ray machines. It is hoped that this unit will use much less electricity than traditional X-ray machines when outfitted with carbon nanotubes that emit electrons for X-ray production. It is hoped that CT and PET/CT can be miniaturized using similar technology. These small imaging machines can be designed to isolate small regions of the body for detailed diagnostic analysis. Ultrasound machines the size of a laptop computer are already being used. Small spectrophotometers are already being devised to replace the tabletop machines used for analyzing blood, spinal fluid, tissues, and urine.

All these improved medical diagnostic technologies will be gaining much benefit from the Internet. Instruments can be designed to communicate to wireless satellite Internet communication systems that convey information about the patient. Information about the patient's medical condition can be uploaded to a medical center or a physician's office. In addition, health records as well as medical reference information can be downloaded to the diagnostic instruments. It is becoming possible for pharmacists to use similar communication capabilities to download a person's genetic information for formulating pharmacogenetic drugs.

Surgical devices are moving away from manual to smart automatic and robotic instruments (Figure 5.4). Currently, these new surgical instruments are large due to limitations in designing the robotic parts. A company called SRI International is already selling a surgical device called the telepresence surgery system (TeSS). The physician performs the operation using computer-mediated surgical tools. These tools improve the physician's accuracy and dexterity. They are especially useful for microsurgery and minimally invasive operations. The telepresence surgery system

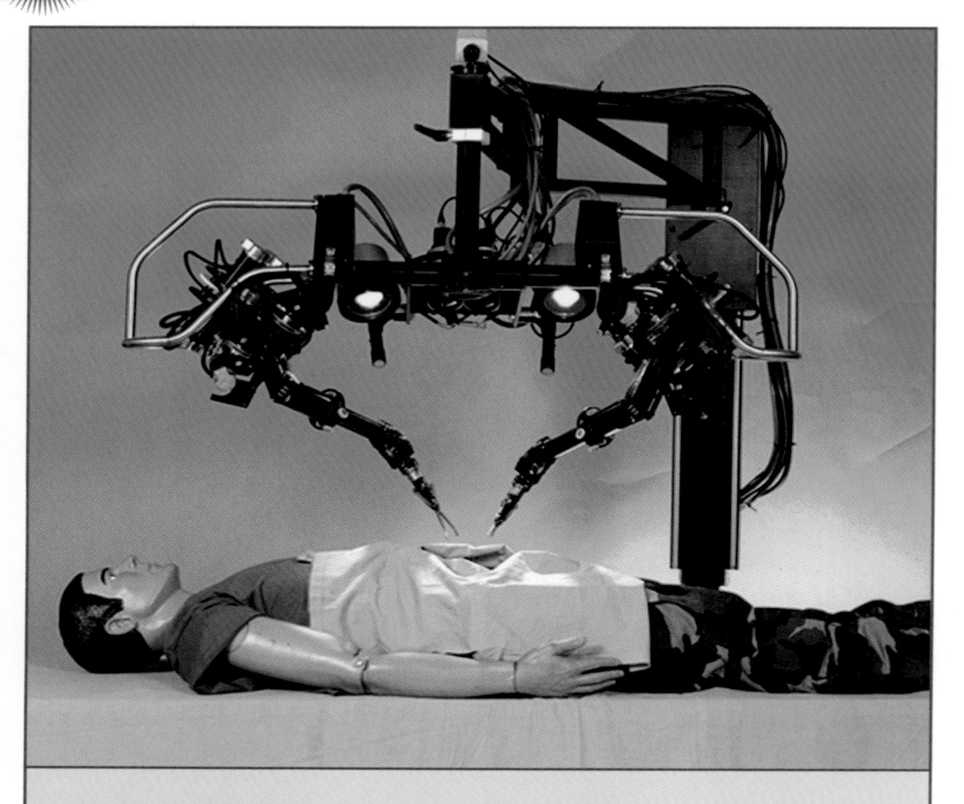

Figure 5.4 Automated or robotic surgical devices, such as the robot pictured above, may someday replace many of the roles of a surgeon.

is similar to the da Vinci Surgical System developed by Intuitive Surgical, Incorporated. Surgical systems such as these will be eventually operated using the same virtual reality controls used in professional flight simulators and advanced computer games. Even the patient's body will be downloaded to the physician as a **hologram** created by feedback from various imaging instruments. A hologram is three-dimensional image of an object exposed to laser light beams.

The scalpels of today are rapidly being transformed into miniature laser scalpels and cryoscalpels. Laser scalpels replace the cutting blade with a beam of laser light that gently burns open a narrow incision. Adjustments of the laser can be made so the scalpel also seals the wounds without the need for sutures. This dual purpose is not available in traditional knife-like scalpels. Laser scalpels are popular for eye surgery and cosmetic surgery because they produce tiny incisions and greatly reduce the chance of scarring. Cryoscalpels use a frozen blade to cut a structure using a procedure called **cryogenic surgery**. This deadens the nerves so little or no pain is felt. The extreme cold also reduces blood flow and consequently minimizes bleeding during delicate operations. It was first tested successfully on delicate tissues found in the kidneys and liver. They can also be used to mend a surgical opening in fragile tissues. Both types of these scalpels can be miniaturized for use in endoscopes and small robotic surgical machines.

Engineers who work in nanotechnology are envisioning handheld devices that can diagnose and treat diseases. These would be very similar to the *Star Trek* tricorder, a handheld medical scanning device. A device similar to the tricorder is already in development. In 1996, Vital Technologies Corporation of Canada developed a scanner called TR-107 Mark 1. It performed environmental monitoring duties that could be adapted to measuring disease conditions in humans. Software developers are developing programs that use handheld computers and pocket PCs for monitoring medical conditions. They will collect data using probes similar to those used in miniaturized diagnostic equipment. It is hoped that these miniscule devices could be outfitted with drug delivery systems and minor surgical tools.[6]

REFERENCES

1. University of Waterloo. "UW Nanotechnology Research Leads to Better Medical Instruments." Available online. URL: http://newsrelease.uwaterloo.ca/news.php?id=4674.
2. T. Gotoh. "Endoscopes with Latest Technology and Concept." *Minimally Invasive Therapy & Allied Technologies.* 12(5) 2004: 222–6.
3. R. Porter. *Cambridge Illustrated History of Medicine.* Cambridge: Cambridge University Press, 1996.
4. J.G. Webster. *Encyclopedia of Medical Devices and Instrumentation.* Hoboken, N.J.: John Wiley & Sons, 2006.
5. P. Aspden, ed. *Medical Innovation in the Changing Healthcare Marketplace: Conference Summary.* Washington, D.C.: National Academy Press, 2002.
6. National Research Council. *Instrumentation for a Better Tomorrow: Proceedings of a Symposium in Honor of Arnold Beckman.* Washington, D.C.: National Academy Press, 2006.

FURTHER READING

Books

Anderson, J., ed. *Devices and Designs: Technology and Medicine in Historical Perspective.* New York: Palgrave Macmillan, 2006.

Bynum, W.F. and R. Porter, eds. *Companion Encyclopedia of the History of Medicine.* London: Routledge, 1997.

Gedeon, Andras. *Science and Technology in Medicine: An Illustrated Account Based on Ninety-Nine Landmark Publications from Five Centuries.* New York: Springer, 2006.

Giddens, S. and O. Giddens. *Future Techniques in Surgery.* New York: Rosen Publishing, 2002.

Porter, R. *The Greatest Benefit to Mankind: A Medical History of Humanity from Antiquity to the Present.* New York: Harper Collins, 1997.

Porter, R., ed. *Cambridge Illustrated History of Medicine.* Cambridge: Cambridge University Press, 1996.

Rutkow, I.M. *Surgery: An Illustrated History.* St. Louis, Miss.: Mosby-Year Book, 1993.

Travers, B. and F.L. Freiman, eds. *Medical Discoveries: Medical Breakthroughs and the People Who Developed Them.* Farmington Hills, Mich.: UXL Publishing, 1997.

Web Sites

Guide to the History of Science
http://www.hssonline.org/guide/

Health and Medicine in the News
http://www.biomed.lib.umn.edu/hmed/

Internet Public Library History of Medicine
http://www.ipl.org/div/subject/browse/hea30.00.00/

Military Medical History
http://www.cs.amedd.army.mil/history/

Science Daily
http://www.sciencedaily.com/news/health_medicine/

GLOSSARY

Absorption The process of a chemical passing through a body surface.

Acellular Refers to the absence of cells.

Adenosine triphosphate (ATP) A molecule that transfers energy from one molecule to another.

Adenovirus A virus that causes respiratory tract infections.

Amino acid One of the building blocks of protein.

Analog A continuous wave of electrical signals that vary in amplitude in response to controlled input.

Angiogenesis factors Chemicals that stimulate the development and growth of blood vessels.

Antibody An immune system chemical that attaches to foreign substances in the body.

Antigen Any substance that can produce an immune response.

Articulated A device that works like a joint.

Autoimmunity A condition in which the body's immune system accidentally attacks particular tissues.

Bioavailability enhancer A chemical that helps drugs enter the blood and the cells causing the ailment.

Biochips Devices composed of cell components or biological molecules attached to glass or plastic surface.

Biocompatibility The inability of a material to cause damage or an immune response in the body.

Bioinformatics The collection, organization, and analysis of large amounts of biological data using networks of computers and databases.

Biomaterial Natural or synthetic substances that are suitable as a substitute for living tissue especially as part of a medical device.

Biomimetics The design of computers and machines by mimicking the natural movements and thought processes of living organisms.

Bionanomachines Microscopic machines produced for bionanotechnology applications.

Bionanotechnology The development of microscope technologies using components of living cells.

Bioprocessing The cultivation of cells or microorganisms for the production of chemicals and cell components.

Biopsy The analysis of living cells removed from the body.

Biotechnology The science of using naturally occurring or modified living organisms for specific medical or technological uses.

Bird flu Also known as avian influenza, it is a type of virus that harms the respiratory system of birds and can be spread to other animals.

Bone marrow A tissue located in the cavities of bones that forms blood cells.

Caesarean delivery Childbirth performed by removing the baby through an incision through the abdominal cavity.

Callus A hard or thickened tissue associated with irritation or a wound.

Capillary A microscopic blood vessel.

Cardiovascular Refers to blood vessels and the heart.

Cartesian Refers to the mathematics of movements and positions.

Cell membrane A lipid and protein covering that encloses the cytoplasm of a cell.

Chemotherapy A treatment of disease using poisonous chemicals that kill cells.

Chloroplast A plant structure that carries out photosynthesis.

Chromatin DNA and associated proteins of a cell.

Chromatography A chemical technique that separates and analyzes mixtures of molecules in solution or in a body fluid.

Chromosome A form of chromatin found in dividing cells.

GLOSSARY

Cilia Short movable hair-like structures found on the surfaces of some types of cells.

Clone Identical cells or organisms derived from a single ancestor.

Cloning The process of producing a clone.

Cochlear A structure in the ear that converts vibrations into sound and then transmits the information to the brain.

Conjugate therapy Treatment with a drug composed of two different therapeutic agents attached to each other.

Cryogenic surgery Surgery performed using a frozen scalpel blade.

Cryopreservation The process of storing cells and whole organisms at low temperatures for long periods of time.

Cybernetic organism Also called a cyborg, is a creature consisting of a mixture of organic and mechanical parts.

Cybernetics The science of creating robots operated by artificial intelligence.

Cyborg A cybernetic organism.

Cystic fibrosis A genetic disease that results in a buildup of mucus in the lungs.

Cytochrome P450 system A chemical pathway that is responsible for the processing of drugs and toxins in the body.

Cytolytic T lymphocytes White blood cells that secret toxins to combat cancer and parasitic infections.

Cytoplasm Cellular material that is within the cell membrane.

Deoxyribonucleic acid (DNA) The chemical that makes up the genetic material of cells.

Diabetes A disease characterized by high levels of sugar in the blood.

Differentiated Cells that carry out specialized functions.

Differentiation factor A chemical that causes a cell to become differentiated to a particular task.

Diffusion The spontaneous spreading of particles.

Dopamine A chemical found in the brain that controls muscle movement.

Doppler A technique that uses sound waves to measure the speed and direction of blood flow.

Dynein A motor protein that can move across the surface of other materials.

Electrocardiogram A diagnostic test that records electrical activity of the heart.

Electroencephalogram A diagnostic test that records electrical activity of the brain.

Electromyogram A diagnostic test that evaluates nerve and muscle function.

Electrophoresis The use of electricity to separate different types of nucleic acids and proteins.

Electrophysiology The measurement of electrical activity of body components.

Embryo An early stage of development before birth.

Emulsion compounds Chemicals that help drugs form a uniform mixture.

Endomembrane system A system of internal membranes within a cell that carries out cell functions.

Endoplasmic reticulum An internal membrane of a cell that carries out specific cell functions.

Endoscope A tubular medical diagnostic instrument that is inserted into the body to view internal components of the body.

Endosymbiont An organism in eukaryotic cells that assists with cell functions.

Endothelial cells Cells that line organs and body cavities.

Endothelin A mitogen involved in skin cell replication.

GLOSSARY

Environomics The science investigating the role of the environment and drugs on the expression of genetic material.

Enzyme A protein that carries out specific chemical reactions.

Epidermal growth factor (EGF) A growth factor involved in the formation of the outermost layer of skin.

Epitope The region of a chemical that binds to the paratope of an antibody.

Ergonomics The process of designing tools so that they are handled using the correct posture and positioning of the body to improve ease of use.

Estrogen A female sex hormone.

Eukaryote A cell that contains a nucleus.

Excipients Substances added to drugs which are required to produce quality tablets, but which do not provide nutritive value.

Expression engineering A method of growing cells so that they produce a particular product.

Ex situ A procedure that is done away from the natural or original location.

Ex vivo A procedure conducted in an artificial environment outside the body.

Fertilized egg An egg that was successfully blended with a sperm cell.

Fetus The later stage of an animal's development before birth.

Fibroblast A cell that grows into connective tissue that supports other tissues.

Flagella Long hair-like projection used for movement in some microorganisms and cells.

Gene A sequence of DNA that represents a fundamental unit of heredity.

Gene therapy The altering of genes in order to treat disease.

Genetic disorder A condition caused partly or completely by a defect in one or more genes.

Genome The total genetic material carried on the chromosomes.

Genomics The study of an organism's complete DNA information.

Germ cell Egg and sperm producing cells found in the gonads or stem cells.

Glial cells Cells that mold and maintain the nervous system.

Global positioning system (GPS) A satellite-based radio positioning system that identifies the location of an object on the Earth.

Golgi apparatus A structure in cells that helps form secretions.

Growth factor A chemical that controls cell division and cell survival.

Growth medium A substance that assists with the growth of cells grown in the laboratory.

Hematocytoblast A pluripotential cell in bone marrow that forms other blood stem cells.

Hematopoiesis The formation of blood cells.

Hematopoietic stem cells Stem cells that form blood.

Hemophilia A genetic disease characterized by uncontrolled bleeding.

Hologram A 3-dimensional image of an object exposed to laser light beams.

Hormone A chemical signal that is usually transported in the blood.

Human Genome Project A government project aimed at sequencing the full complement of human DNA.

Huntington's disease A genetic disorder that causes the loss of nerve cells in a specific part of the brain.

GLOSSARY

Hydrocarbon polymer A molecule consisting only of chains of carbon and hydrogen.

Hydrogen bond A weak bond formed by electrical charges between hydrogen and other atoms.

Immune response A set of reactions the body uses to attack and remove foreign substances that enter the body.

Immunization A process by which protection from an infectious disease or cancer is administered.

Immunoglobulin A protein used to battle foreign substances.

Immunoinformatics The science of investigating how the immune system functions in response to disease and drugs.

Incubator A chamber used to grow cells or organisms under precise environmental conditions.

Infrared light Light beyond the visible range; it is capable of heating objects.

In situ A procedure that is done in the natural or original location.

Insulin A hormone that helps adjust blood sugar.

Insulin-like growth factor-I (IGF-I) A chemical that assists other factors with the differentiation of stem cells.

Intravenous The administration of a drug or fluid directly into a vein.

In vitro Refers to culturing organisms or conducting experiments under laboratory conditions.

In vivo A procedure carried out within the body.

Leukemia A cancer of the bone marrow that produces white blood cells.

Ligand A chemical attached to another to impart specific characteristics.

Ligase An enzyme used to combine the cut ends of DNA fragments.

Liquid vehicles Fluids that are mixed with the drug to make it uniform and palatable.

Lubricants Chemicals added to drugs that help the drug flow through the machinery and equipment used to mix and mold the drugs.

Lupus Also known as systemic lupus erythematosus (SLE), it is a disease in which the immune system attacks many body components especially the skin.

Lymph nodes Structures that help the body fight disease and assist with the repair of tissues.

Lysosome A structure in a cell capable of dissolving material that enters the cell.

Magnetic resonance imaging (MRI) An imaging technology that uses powerful magnets to measure chemical characteristics of tissues.

Mammary gland A milk-producing gland found in mammals.

Matrix A framework that supports a structure.

Medical imaging A technique that permits physicians to evaluate areas of the body that are not normally visible.

Mesenchymal stem cells A collection of embryonic cells related to stem cells.

Metabolism A series of chemical reactions that carry out the functions of a cell and an organism.

Metabolomics A study of how genes affect the metabolism of a cell.

Microarray A biochip for studying how large numbers of genes interact with each other.

Microemulsion compounds Chemicals that help drugs form a uniform mixture.

Microtubules Hollow microscopic tubes made of protein or carbon atoms.

GLOSSARY

Mitochondria Endosymbiont organelles that produce energy for the cell.

Mitogen A signal that stimulates the replication of cells.

Molecular biology The scientific study of the structure and function of biologically important molecules.

Multipotential A stem cell that can develop into a variety of different cell types.

Nanobiosensors Cell-driven microscopic instruments for detecting chemicals.

Nanobiotechnology Devices built using biological molecules, cells, or components of cells.

Nerve cell A cell that receives and sends messages from the body to the brain and back to the body.

Nuclear membrane A covering that forms the surface of the nucleus.

Nucleus A structure in eukaryotic cells that contains the genetic material.

Oncogene A cancer-promoting gene.

Organ A specialized structure in an organism that carries out particular body functions.

Organelle A structure in eukaryotic cells that carries out particular cell functions.

Papyrus A sheet of paper made from a water reed that was abundant in ancient Egypt.

Parasites Organisms that live in or on the living tissue of a host organism at the expense of the host.

Paratope A region of an antibody that attaches to a chemical.

Parkinson's disease A progressive degenerative brain disorder.

Patent A government grant giving an inventor the exclusive right to make or sell an invention for a term of years.

Pharmaceutical Any chemical that has medical value as a drug.

Pharmacodynamics The processes involved in a drug's effect on a cell, tissue, organ, or the whole body.

Pharmacogenetics The study of how a person's particular genetic make-up affects their response to therapeutic treatments.

Pharmacokinetics The study of drug metabolism related to the time required for absorption, duration of action, distribution in the body, and removal from the body.

Photosynthesis A plant process that uses energy from sunlight to convert water and carbon dioxide into carbohydrates and oxygen.

Physiomics The study of the genetics of metabolic functions in the body.

Placenta An organ that nourishes the developing fetus in the uterus.

Plasma An extremely hot gas that is composed of charged particles.

Plasmid A small circular piece of bacterial DNA.

Plasticity The ability of a stem cell from an adult tissue to produce the differentiated cell types of another tissue.

Pluripotential A stem cell that can develop into nearly all of the different cell types.

Positron A positively charged subatomic particle.

Preventative medicine A branch of medicine that includes practices that help people avoid disease and the promotion of health.

Prokaryote A primitive cell without a nucleus that makes up bacteria.

Prostheses Another term for prosthetic devices.

Prosthetic device A device used as a substitute for a missing body part.

Protists Microscopic organisms that have a nucleus and are generally composed of a single cell.

Protocell An artificial cell that carries out some of the functions of living organisms.

Receptor A protein of the surface of cells that is able to detect specific types of chemicals.

Recombinant DNA A novel sequence of genetic material that is formed by combining pieces of DNA from different organisms or cells.

Red blood cells Blood cells that carry oxygen throughout the body.

Regeneration The growth of a new or lost tissue.

Rejection An immune system response in which the body destroys a foreign material in the body such as a transplanted organ or medical device.

Restriction enzymes Bacterial enzymes that cut DNA at very specific locations.

Retina The inner layer of the eye containing nerve cells that detect light.

Retrovirus A type of virus that has RNA instead of DNA as its genetic material.

Rheumatoid arthritis A disease that causes inflammation in the lining of the joints.

RNA Ribonucleic acid, which is a chemical similar to DNA from which proteins are made.

RNA interference A treatment that works by using specially made RNA strands that inhibit the expression of particular genes.

Scaffold A framework that supports a structure.

Secretion A chemical with a particular function that is released by a cell into the body or the environment.

Servo A small motor driven device that produces specific angular positions in robotic devices.

Severe combined immune deficiency (SCID) A genetic disorder that disables the immune system making it impossible for the body to fight off disease.

Sickle cell anemia An inherited disease of red blood cells that limits the blood's ability to carry oxygen.

Single nucleotide polymorphism (SNP) A common mutation consisting of a change at a single base in the DNA.

Somatic cell A cell that makes up the body and is not involved in sexual reproduction.

Somatostatin A protein that stimulates the production of human growth hormone.

Spectrophotometer An instrument used to measure the amount of light reflected from a specimen when illuminated by a controlled light source.

Stem cell A cell from which other types of cells of the body can develop.

Syringe A small hollow needle or tube used for injecting or withdrawing liquids.

Testosterone A male sex hormone.

Therapeutic A chemical or procedure used as a medical treatment.

Therapeutic cloning The use of somatic cell nuclear transfer to produce stem cells that differentiate into tissues.

Tissue A group of similar cells that work together to carry out a particular function.

Tomography An X-ray technique that produces detailed cross-sections of the body.

Tonometry A measurement of the pressure or swelling of a body part.

GLOSSARY

Totipotential The ability to develop into any cell type in an organism.

Transdermal Passing through the skin into the body.

Transducer Any device that converts one form of energy into another form of energy.

Transgenic An organism that has genes from another organism inserted into its DNA using recombinant DNA techniques.

Transplantation The replacement of tissue with tissue from another person.

Transport vesicle A small organelle that moves material around the cell.

Trepanation An ancient form of surgery where a hole is drilled or scraped into the skull. It is often called trephinning or trepanning.

Turing machine A simple computational device designed to investigate the simplest path to determine a mathematical calculation.

Ultrasound A medical imaging method that uses high energy sound waves to outline a part of the body.

Umbilical cord A tube, containing blood vessels, that connects the developing fetus to the mother.

Unipotential A stem cell capable of forming one type of cell.

Vaccination The process by which a person's immune system is induced to develop protection from a particular disease.

Vesicle A small sack-like organelle in the cytoplasm of a eukaryotic cell.

Virus A small particle that infects cells.

White blood cell A group of blood cells involved in the immune response.

Xeroderma pigmentosum An inherited disorder that causes extreme sensitivity to sunlight and early onset of skin cancers.

X-ray A form of radiation used for imaging internal structures of the body.

PICTURE CREDITS

1: © UMW / Visuals Unlimited
4: © Infobase Publishing
8: © Infobase Publishing
10: © Dr. Yorgos Nikas / Photo Researchers, Inc.
14: © Associated Press
18: © Associated Press
20: © Infobase Publishing
27: U.S. Department of Energy Genomics: GTL Program, http://genomicsgtl.energy.gov
29: © Associated Press
33: © Time Life Pictures/Getty Images
34: © Infobase Publishing
41: © Associated Press
46: Courtesy Dr. Charles Vacari
55: © AP Photo/Paul Sakuma
57: © Associated Press
61: © Associated Press
71: © Associated Press
75: © Associated Press
79: U.S. Department of Energy Human Genome Program, http://www.ornl.gov/hgmis
87: © Associated Press
92: © Damien Lovegrove / Photo Researchers, Inc.
97: © YOSHIKAZU TSUNO/ AFP/Getty Images
99: Courtesy of PEMED, Denver, Colorado, www.pemed.com
100: © Associated Press
105: © Associated Press
114: Courtesy SRI International

Cover: © Pascal Goetgheluck / Photo Researchers, Inc.

INDEX

Acellular replacement prosthesis
  uses, 43, 45–46
Adelman, Leonard, 66
Adenosine triphosphate (ATP), 65
Adenovirus, 38
Aiken, Howard, 58
Alzheimer's disease, 23
Analog circuits, 72
Anderson, W. French
  research, 35–36
Angiogenesis factors, 41
Animatronics, 68
Antibodies
  artificial, 90–91
  function, 89
  paratope, 90
  production, 13
Antigens, 13
Artificial neural networks
  uses, 43, 48–49
Asimov, Isaac, 59
  The Bicentennial Man, 70
Asthma, 86
ATP. *See* Adenosine triphosphate

BDAS. *See* Boston Digital Arm System
Bellamkona, Ravi
  research, 44
Berg, Paul
  research, 31–32
Bicentennial Man, The (Asimov), 70
Biochips, 48, 86
Bioinformatics, 81
Biomaterials
  diagnostic devices, 39
  medical devices, 39–40, 74
  medical imaging, 39
  natural, 40, 43, 48
  prostheses, 39
  synthetic, 40–41, 43, 45, 48
Biomimetics, 72
Bionanomachines
  building materials, 63, 65
Bionanotechnology
  defined, 60–63
  devices, 62–63, 65
  research, 60–63, 65–67
  specialty fields of, 62–63
  uses, 59, 63, 65–67
Bioprocessing, 13
Biopsy, 112
Biosensors, 47–48
Biotechnology
  advances in, 80
  applications, 32
  divisions, 39–40
  modern, 31, 86
  research, 6, 15, 21, 23
Blaese, Michael, 35
Blood disorders
  and stem cell treatment, 14–17, 43
Blumenkranz, Mark
  research, 44
Boland, Thomas, 28
Bone marrow
  cells in, 6–7, 11
Boston Digital Arm System (BDAS), 73
Boyer, Herbert
  research, 31–32
Brain
  damage and disability, 67
  functional systems, 49, 72, 74
Burns
  stem cell treatment, 14–15, 23, 28
  tissue engineering to treat, 46
Bush, George W.
  stem cell research, 43

Caesarian deliveries, 101
Cancer
  promoting gene, 33
  research, 47
  treatments, 15–16, 30, 37–38, 47, 58, 82–83, 91
  types, 82
  vaccines, 88, 90–93
Capek, Karel
  Rossum's Universal Robots, 59
Cardiovascular system
  diseases, 15
  heart valve repair, 56
  prosthetics, 73–74
  restoration, 14–15, 43
CardioWest temporary Total Artificial Heart (TAH-t), 73
CAT scan. *See* Computed tomography
Cell
  basics, 3–6
  differentiation, 6, 19
  functions, 3, 63, 107
  living machines, 47–48
  major divisions, 3, 43
  metabolism, 5
  product delivery systems, 43
  reproduction, 82
  structures, 106
  types, 7, 9, 21, 28, 30, 36, 42, 44, 66–67, 83
Cell membrane
  function, 3
Cellular prosthesis
  uses, 43–45
Chemotherapy
  and cancer treatment, 15–16, 82–83, 91
  and parasitic diseases, 82
Chicken pox
  vaccines, 89–90
Chromatin, 5
Chromatography, 108
Chromosome, 5, 83
Cloning
  animal, 17, 19–20, 22
  defined, 17, 19–21
  ethical issues, 19, 23
  humans, 19

INDEX

medical applications, 3, 21
plant, 19
reproductive, 19
research, 17, 19–23
uses, 21–23
Cohen, Stanley
research, 31–32
Computed tomography (CAT scan), 107, 113
Conjugate therapy, 91
Cryogenic surgery, 115
Cryopreservation, 13
Culver, Kenneth, 35
Cybernetics
development, 59
Cyborgs, 48, 70
Cystic fibrosis, 38
Cytochrome P450 system, 84–85
Cytometers, 108
Cytoplasm
function, 3–4

da Vinci Surgical System
development, 56, 114
Delivery systems, 115
infusion, 103–104, 110
internal, 111
needleless injection, 103–104, 110–111
syringe injection, 103–105, 111
Deoxyribonucleic acid (DNA)
computer, 66–67
damage, 48
function, 5, 63, 107
information, 81, 86
nanoprinting, 28
neural networks, 62
plasmid, 31–21
recombinant, 31, 33
replication, 32
research, 13, 19, 21–22, 30, 37
Descartes, René, 68
DeSilva, Ashanti
gene therapy for SCID, 35–37
Diabetes
artificial pancreas to treat, 75
and living insulin injectors, 47
treatment with stem cells, 14, 16
Diphtheria, 90
DNA. *See* Deoxyribonucleic acid
Dopamine
and the treatment of Parkinson's disease, 22–23
Drexler, Kim Eric
research, 60–61

Ebers papyrus, 30
ECG. *See* Electrocardiograms
Eckert, J. Presper, 58
EEG. *See* Electroencephalograms
Einthoven, Willem, 102
Electrical Numerical Integrator and Calculator (ENIAC 1), 58
Electrocardiograms (ECG), 76, 102, 106
Electroencephalograms (EEG), 76, 102, 106
Electromyogram (EMG), 102, 106
Electronic indentation tonometry, 112–113
Electrophoresis, 66
Electrophysiology, 106
Embryo
development, 6, 8–9, 22
splitting, 17, 19
stem cells, 6, 9, 11–13
EMG. *See* Electromyogram
Endoscope, 98, 115
research, 99, 101
ENIAC 1. *See* Electrical Numerical Integrator and Calculator
Environmomics, 81
Enzymes
bacterial, 32
dynein, 65
ligase, 31–32
restriction, 31
Ergonomics, 109
Eukaryotes
organelles, 4–5
Expression engineering, 13
Ex vivo technique, 36–38

FDA. *See* Food and Drug Administration
Feynman, Richard, 60
Fishman, Harvey
research, 43–44
Food and Drug Administration (FDA), 36, 40
Fox, Michael J., 11
Fuller, Buckminster
research, 61–62

Gelsinger, Jesse, 35–36
Gene therapy
defined, 31–35, 67
public controversy, 36
research, 16, 31–39, 84
uses, 31, 35–39
Genetic disorders
testing, 12–13, 85–86
treatment, 3, 30–31, 37–38, 67
Genetic engineering
and cloning, 19–20
trials, 31–32, 60, 110
Genome, 3
altering, 31
control, 7
function, 5, 7, 83
identification, 28, 83, 86
modification, 13, 17, 32, 35, 37, 39
Genomics, 81, 83
German measles
vaccines, 89–90
Germ cell therapy, 38–39
Glial cells, 44

Global Positioning system (GPS), 65–66
Goodman, Howard, 32
Government
  and gene therapy, 36
  grants, 33, 83
  safety guidelines, 110
  and stem cell research, 9, 43
GPS. *See* Global Positioning system
Grasso, Frank, 72
Growth factors
  brain-derived neurotrophic, 2
  and mitogens, 7, 9
  research, 2
  and stem cell differentiation, 7–9, 12, 15, 21–22, 41–42
Gurdon, John, 17

Harvard Mark I computer, 58
Hematopoiesis, 14
Hemophilia, 38
Hepatitis A
  vaccines, 89
Holwill, Michael E.J., 65
Hopper, Grace, 58
Hormones
  growth, 32, 37
  plant, 17
  testing, 108
  types, 8–9, 16, 45
HUGO. *See* Human Genome Organization
Human Genome Organization (HUGO), 31
Human Genome Project
  development, 82–83
Human papillomavirus, 92–93
Huntington's disease, 38

Illmensee, Karl, 17
Immune system
  and cellular prosthesis, 45
  disables, 35
  enhancement, 14, 16, 50, 89–90, 93
  epitope's effects on, 80–81
  functions, 80, 82, 90, 92
  research, 82
  response, 13–14, 35, 40, 80–81, 90–91, 93
  structures of, 45
  and vaccination, 82
Immunoglobulin, 89
Immunoinformatics
  research, 80–81
Infectious
  agents, 83, 90, 93
  diseases, 86, 88–89, 93
Influenza
  avian, 91–92
  vaccines, 89–90
International Immunoinformatics Symposium, 80
In vivo technique, 37–38

Jarvik, Robert K., 74

Kroto, Harry, 62

Leder, Philip, 33
Leukemia, 36
Living machines, 47–48
Lupus, 16

Magnetic resonance imaging (MRI), 106
Marmor, Michael
  research, 44
Mauchly, John, 58
McCulloch, Warren, 49
Measles
  vaccines, 89–90
Medical instrumentation
  delivery, 103–105, 110–112, 115
  design and guidelines, 103
  devices, 39–40, 98–115
  early, 98, 101–102
  future improvements to, 98–99, 101, 110–115
  imaging devices, 39, 56–57, 76, 103, 106–108, 112–115
  research, 98–99, 101
  surgical, 98–99, 101–104, 109, 113–115
Metabolomics, 81, 83
Microarrays, 28, 86
Microscope, 102
  fluorescent, 106–107
Mironov, Vladimir, 28
Molecular biology, 110
Monitored Automatic Pill Dispenser, 110–111
Montagu, Mary Wortley, 88
MRI. *See* Magnetic resonance imaging
Multipotential stem cells, 6
  cultures, 11
  and formation of red blood cells, 7
  sources of, 10–11, 21–23
Mumps, 90

Nanobiosensors, 50
Nanobiotechnology
  components, 59
  defined, 43
  development, 60, 62
  practicality of, 61
  research, 60–63, 98, 101, 115
  strategies, 49–50
Nano Ethics Conference, 62–63
National Academy of Sciences, 83
National Aeronautics and Space Administration, 60, 112
National Institutes of Health, 12, 82
National Spinal Cord Injury Statistical Center, 2

INDEX

Nervous system
- disorders, 2, 11
- functions of, 49
- injuries, 11

Neurons
- death, 2
- functions of, 49, 72, 88, 112
- inflammation of, 89
- regeneration, 2, 14

Oncogene, 33

Organ
- and cell production, 11
- and cloning, 23
- development, 6
- modeling system, 43, 45
- printing, 28
- regeneration of damaged, 3, 9, 13, 30, 47
- transplantation, 14, 23

Ornithine transcarbamylase disorder, 35

Palanker, Daniel
- research, 43–44

Parkinson's disease
- treatments, 22, 67

Pertussis, 90

PET scan. *See* Positron emission tomography

Pharmaceuticals
- research, 6, 12, 23, 82, 84

Pharmacogenetics
- and clinical testing, 80, 86, 88
- defined, 83–85
- pharmacodynamics, 83
- pharmacokinetics, 83
- principles of, 84
- research, 80, 83–86, 88
- synopsis of, 85
- uses, 85–86, 88

Photosynthesis, 5

Physiomics, 81

Pits, Walter, 49

Pluripotential stem cells, 6
- sources, 10–11, 21–22
- and tissue development, 6

Poliomyelitis
- vaccines, 89–90

Positron emission tomography (PET scan), 107, 113

Pravaz, Charles Gabriel, 105

Precision vaccines
- defined, 88–90
- development of, 90
- research, 88–93
- uses, 91–93

Prokaryotes
- structures, 3–5

Prosthesis
- acellular replacement, 43, 45–46
- and amputation, 30
- cardiac, 73
- cellular, 43–45
- modern, 31, 39
- ocular, 74
- robotics, 72–74

Protocells
- discovery, 66–67

Radiography
- fluoroscopy, 107
- improvements, 107–108, 113

RBS. *See* Robotics control system

Reeve, Christopher, 2, 11

Reproductive cloning, 19

Research
- on bionanotechnology, 60–63, 65–67
- on cloning, 17, 19–23
- on gene therapy, 16, 31–39, 84
- on the immune system, 82
- on immunoinformatics, 80–81
- on instrumentation, 98–99, 101
- on nanotechnology, 60–63, 98, 101, 115
- on pharmacogenetics, 80, 82, 83–86, 88
- on robotics, 57–58, 75–76
- on stem cells, 2–3, 6, 9–16, 23, 39, 43, 47
- on targeted medicine, 81–83, 86, 88
- on tissue engineering, 28, 42–45, 47–50
- on vaccines, 88–93

Retrovirus, 37

Rheumatoid arthritis, 16
- causes, 80

RNA interference, 86

Robotics
- articulated, 68–69
- axonemal, 65–66
- cameras, 70, 76
- cartesian, 68–69
- consumer, 68
- defined, 67–70
- microscopic, 60
- parallel, 69
- prosthetics, 72–74
- research, 57–58, 72–75
- self-replicating, 63
- sensors, 70, 76
- surgery, 56–57, 74–76, 115
- uses, 58–59, 67–68, 70, 72–76

Robotics control system (RBS), 69–70

Roentgen, Wilhelm Conrad, 107

Rossum's Universal Robots (Capek), 59

Rutter, Bill, 32

Salk, Jonas, 89

SCARA. *See* Selective Compliant Articulated/Assembly Robot Arm robots

SCID. *See* Severe combined immune deficiency

Selective Compliant Articulated/Assembly Robot Arm robots (SCARA), 69
Severe combined immune deficiency (SCID)
treatment, 35–36
Sickle cell anemia, 38
Smalley, Richard, 62
Smallpox
vaccines, 88, 90
Smith, Hamilton
research, 31–32
Somatic cell nuclear transfer, 19
Spectrophotometer, 108
miniature, 113
Spemann, Hans, 6
Spinal cord
cells, 2–3
Spinal cord injuries
research, 2–3
and stem cell, 2
Stellacci, Francesco, 28
Stem cells
adult tissues, 9, 11–12
control of function, 7–9, 12
defined, 5
embryonic, 2, 9, 11–13
ethical issues, 3, 9, 11, 23
and health risks, 3
lineages, 12, 21
medical applications, 9, 12–17
preparation, 12–13
regeneration, 14, 21, 42
replication, 9
research, 2–3, 6, 9–16, 23, 39, 43, 47
sources, 9–11, 22
types, 6–7, 10–11, 13, 15–17, 21–23
umbilical, 9–12
uses, 11–17
Stethoscope, 102
Stewart, Timothy, 33
SynCardia Systems, 73

TAH-t. *See* CardioWest temporary Total Artificial Heart
Taniguchi, Norio, 60
Targeted medicine
bioavailability enhancers, 84–85
categories, 82, 88–90
emulsion, 84–85
excipients, 84
liquid vehicles, 84
lubricants, 84–85
microemulsion compounds, 84–85
research, 81–83, 86, 88
Taylor, Helen C., 65
Telepresence surgery system (TeSS), 113–114
Tetanus
vaccines, 89–90
Tissue, 106
artificial, 28, 39, 43–48, 50
regeneration, 3, 6, 9, 11, 13–14, 23, 40, 42
repair, 30, 39, 41–42
scaffolds, 29, 39–40, 42, 45
Tissue engineering, 31
applications, 42–43
defined, 39–42
importance of, 42
nature of, 112
research, 28, 42–45, 47–50
steps, 40
uses, 42–50
Totipotential stem cells
and cell production, 6
sources, 11
Transgeneic, 33
Transplantation
and cloning, 21
rejection, 16
and stem cell research, 14, 16
Trepanation, 101
Turing, Alan Mathison, 66
Turing machine, 66

Ultrasound, 107
Doppler, 76, 108
improvements, 112–113
Umbilical cord
stem cells, 9–11
UNECE. *See* United Nations Economic Commission for Europe
Unipotential stem cells, 6
sources, 11, 23
tissue specific, 7
United Nations Economic Commission for Europe (UNECE), 68

Vaccines, 3
active, 89–90
defined, 82
delivery, 93, 104
early, 88–91
functions, 89
passive, 89–90
precision, 88–93
research, 88–93
side effects, 90
Venter, Craig, 83
Verfaillie, Catherine
research, 6
Viruses
artificial, 67
removal, 50
types, 37–38, 89, 92–93
von Basch, Samuel, 102

Wilcox, Kent
research, 31–32
Willadsen, Steen, 19
Woo, Alexander, 105

Xeroderma pigmentosum, 38

Yeow, John
research, 98–99, 101

ABOUT THE AUTHOR

BRIAN R. SHMAEFSKY, Ph.D., is the author of several books about biotechnology and human disease. He has written more than 100 articles on college science teaching. Dr. Shmaefsky also consults on biotechnology development and performs biotechnology teacher training for the Biotechnology Institute. His research interests include the effects of pollutants on animal and human development.

ABOUT THE EDITOR

YAEL CALHOUN is a graduate of Brown University. She has an M.A. in education and an M.S. in natural resources science. Currently, she writes books and teaches environmental biology at Westminster College in Salt Lake City. She lives with her family at the foot of the Rocky Mountains in Utah.